Lecture Notes of the Institute for Computer Sciences, Social-Informatics and Telecommunications Engineering 14

Sanmay Das   Michael Ostrovsky
David M. Pennock   Boleslaw Szymanski (Eds.)

Auctions, Market Mechanisms and Their Applications

First International ICST Conference, AMMA 2009
Boston, MA, USA, May 8-9, 2009
Revised Selected Papers

 Springer

Volume Editors

Sanmay Das
Boleslaw Szymanski
Rensselaer Polytechnic Institute
110 8th St, Troy, NY 12180 USA
E-mail: {sanmay; szymansk}@cs.rpi.edu

Michael Ostrovsky
Stanford University Graduate School of Business
518 Memorial Way, Stanford, CA 94305 USA
E-mail: ostrovsky@gsb.stanford.edu

David M. Pennock
Yahoo! Research, 111 West 40th Street
17th floor, New York, NY 10018, USA
E-mail: pennockd@yahoo-inc.com

Library of Congress Control Number: 2009932900

CR Subject Classification (1998): J.1, K.1, C.2.4, C.3, H.2.8, H.4

ISSN   1867-8211
ISBN-10   3-642-03820-4 Springer Berlin Heidelberg New York
ISBN-13   978-3-642-03820-4 Springer Berlin Heidelberg New York

springer.com

© ICST Institute for Computer Science, Social Informatics and Telecommunications Engineering 2009
Printed in Germany

Typesetting: Camera-ready by author, data conversion by Scientific Publishing Services, Chennai, India
Printed on acid-free paper   SPIN: 12731527   06/3180   5 4 3 2 1 0

Preface

These proceedings present the technical contributions to the First Conference on Auctions, Market Mechanisms, and Their Applications (AMMA), held May 8-9, 2009 in Boston, Massachusetts, USA. The conference was devoted to issues that arise in all stages of deploying a market mechanism to solve a problem, including theoretical and empirical examinations. In addition to more traditional academic papers, the conference placed emphasis on experiences from the real world, including case studies and new applications.

The main goal of AMMA was to explore the synergy required for good mechanism design. This includes an understanding of the economic and game-theoretic issues, the ability to design protocols and algorithms for realizing desired outcomes, and the knowledge of specific institutional details that are important in practical applications. We were lucky enough to attract papers and talks from economists and computer scientists, theorists and empiricists, academics and practitioners. The program, as reflected in these proceedings, ranged from fundamental theory on auctions and markets to empirical design and analysis of matching mechanisms, peer-to-peer-systems, and prediction markets.

The call for papers attracted submissions from academia and industry around the world, including Asia, North America, Europe, and the Middle East. Each paper was reviewed by at least three Program Committee members on the basis of scientific novelty, technical quality, and importance to the field. The Program Committee selected 18 papers for presentation at the conference out of 38 submissions, most of which are published in the proceedings. At the authors' request, only abstracts of some papers are included in these proceedings. This accommodates the practices of fields outside of computer science in which journal rather than conference publishing is the norm and conference publishing sometimes precludes journal publishing. It is expected that many of these papers will appear in a more polished and complete form in scientific journals in the future.

Putting together AMMA 2009 was a team effort. We thank the authors for providing the content of the program. We are grateful to the Program Committee and the external reviewers, who worked hard reviewing and providing feedback for authors. We thank Local Chair Peter Coles for making sure everything at the venue went off without a hitch. We thank Conference Coordinator Robert Varga, Finance Chair Gabriella Pal, Publicity Chair Sebastien Lahaie, and Webmaster Micha l Karpowicz. Finally, we thank the conference sponsor, the Institute for Computer Sciences, Social-Informatics and Telecommunications Engineering (ICST).

June 2009

Michael Ostrovsky
David Pennock
Boleslaw Szymanski

Organization

Steering Committee Co-chairs

Imrich Chlamtac   U. Trento and Create-Net, Italy
Boleslaw Szymanski   Rensselaer Polytechnic Institute, USA

Steering Committee Members

Peter Coles   Harvard Business School, USA
Sanmay Das   Rensselaer Polytechnic Institute, USA
Michael Ostrovsky   Stanford GSB, USA
David Pennock   Yahoo! Research, USA
Vincent Conitzer   Duke University, USA

General Co-chairs

David Pennock   Yahoo! Research, USA
Boleslaw Szymanski   Rensselaer Polytechnic Institute, USA

Technical Program Chairs

Sanmay Das   Rensselaer Polytechnic Institute, USA
Michael Ostrovsky   Stanford GSB, USA

Local Chair

Peter Coles   Harvard Business School, USA

Conference Coordinator

Robert Varga   ICST

Finance Chair

Gabriella Pal   ICST

Publicity Chair

Sebastien Lahaie   Yahoo! Research, USA

Webmaster

Michal Karpowicz   Warsaw University of Technology and NASK,
                                          Poland

Technical Program Committee

Atila Abdulkadiroglu	Duke University, USA
Elliot Anshelevich	Rensselaer Polytechnic Institute, USA
Itai Ashlagi	Harvard Business School, USA
Imrich Chlamtac	U. Trento and Create-Net, Italy
Peter Coles	Harvard Business School, USA
Vincent Conitzer	Duke University, USA
Sanmay Das	Rensselaer Polytechnic Institute, USA
Mathijs de Weerdt	Delft University of Technology, The Netherlands
Erol Gelenbe	Imperial College London, UK
Enrico Gerding	University of Southampton, UK
Nicole Immorlica	Northwestern University, USA
Kamal Jain	Microsoft Research
Sven Koenig	University of Southern California, USA
Sebastien Lahaie	Yahoo! Research, USA
Kevin Leyton-Brown	University of British Columbia, Canada
Robin Lee	NYU Stern and Yahoo! Research, USA
Malik Magdon-Ismail	Rensselaer Polytechnic Institute, USA
Mohammad Mahdian	Yahoo! Research, USA
Ewa Niewiadomska-Szynkiewicz	Warsaw University of Technology / NASK Poland
Michael Ostrovsky	Stanford GSB, USA
David Pennock	Yahoo! Research, USA
Ariel Procaccia	Microsoft Israel R&D Center, Israel
Juan Rodriguez-Aguilar	IIIA, CSIC Barcelona, Spain
David Sarne	Bar-Ilan University, Israel
Boleslaw Szymanski	Rensselaer Polytechnic Institute, USA
Ioannis Vetsikas	University of Southampton, UK

Table of Contents

Effects of Suboptimal Bidding in Combinatorial Auctions   1
    Stefan Schneider, Pasha Shabalin, and Martin Bichler

Using Prediction Markets to Track Information Flows: Evidence from
Google   3
    Bo Cowgill, Justin Wolfers, and Eric Zitzewitz

A Copula Function Approach to Infer Correlation in Prediction
Markets   4
    Agostino Capponi and Umberto Cherubini

Manipulating Scrip Systems: Sybils and Collusion   13
    Ian A. Kash, Eric J. Friedman, and Joseph Y. Halpern

A Centralized Auction Mechanism for the Disability and Survivors
Insurance in Chile   25
    Gonzalo Reyes H.

Information Feedback and Efficiency in Multiattribute Double
Auctions   26
    Kevin M. Lochner and Michael P. Wellman

Impact of Misalignment of Trading Agent Strategy across Multiple
Markets   40
    Jung-woo Sohn, Sooyeon Lee, and Tracy Mullen

Market Design for a P2P Backup System (Extended Abstract)   55
    Sven Seuken, Denis Charles, Max Chickering, and Sidd Puri

School Choice: The Case for the Boston Mechanism   58
    Antonio Miralles

Turing Trade: A Hybrid of a Turing Test and a Prediction Market   61
    Joseph Farfel and Vincent Conitzer

A Market-Based Approach to Multi-factory Scheduling   74
    Perukrishnen Vytelingum, Alex Rogers, Douglas K. Macbeth,
    Partha Dutta, Armin Stranjak, and Nicholas R. Jennings

Auctions with Dynamic Populations: Efficiency and Revenue
Maximization (Extended Abstract)   87
    Maher Said

Revenue Submodularity   89
    Shaddin Dughmi, Tim Roughgarden, and Mukund Sundararajan

Fair Package Assignment.. 92
   Sébastien Lahaie and David C. Parkes

Solving Winner Determination Problems for Auctions with Economies
of Scope and Scale 93
   Martin Bichler, Stefan Schneider, Kemal Guler, and Mehmet Sayal

Running Out of Numbers: Scarcity of IP Addresses and What to Do
about It 95
   Benjamin Edelman

Author Index.. 107

Effects of Suboptimal Bidding in Combinatorial Auctions

Stefan Schneider, Pasha Shabalin, and Martin Bichler

Technische Universität München, Germany
{schneider,shabalin,bichler}@in.tum.de
http://ibis.in.tum.de/research/ca/index.htm

Abstract. Though the VCG auction assumes a central place in the mechanism design literature, there are a number of reasons for favoring iterative combinatorial auction designs. Several promising ascending auction formats have been developed throughout the past few years based on primal-dual and subgradient algorithms and linear programming theory. Prices are interpreted as a feasible dual solution and the provisional allocation is interpreted as a feasible primal solution. iBundle(3) (Parkes and Ungar 2000), dVSV (de Vries et al. 2007) and the Ascending Proxy auction (Ausubel and Milgrom 2002) result in VCG payoffs when the coalitional value function satisfies the buyer submodularity condition and bidders bid straightforward, which is an ex-post Nash equilibrium in that case. iBEA and CreditDebit auctions (Mishra and Parkes 2007) do not even require the buyer submodularity condition and achieve the same properties for general valuations. In many situations, however, one cannot assume bidders to bid straightforward and it is not clear from the theory how these non-linear personalized price auctions (NLPPAs) perform in this case. Robustness of auctions with respect to different bidding behavior is therefore a critical issue for any application. We have conducted a large number of computational experiments to analyze the performance of NLPPA designs with respect to different bidding strategies and different valuation models. We compare the results of NLPPAs to those of the VCG auction and those of iterative combinatorial auctions with approximate linear prices, such as ALPS (Bichler et al. 2009) and the Combinatorial Clock auction (Porter et al. 2003).

While NLPPAs performed very well in case of straightforward bidding, we could observe problems with respect to auctioneer revenue, efficiency, and speed of convergence in case of non-best-response bidding behavior. Under suboptimal bidding strategies, dVSV and CreditDebit auctions have a significantly lower efficiency than iBundle(3).

In contrast, linear price combinatorial auctions were robust against all strategies, except collusive behavior which makes it very difficult for any auctioneer to select an efficient solution in general. While our results achieve high efficiency values on average, one can easily construct examples, where linear price CAs cannot be efficient (Bichler et al. 2009). Linear-price designs suffer from the fact that they cannot be 100% efficient, but they have shown to be robust against many strategies and bear a few advantages:

S. Das et al. (Eds.): Amma 2009, LNICST 14, pp. 1–2, 2009.

- Only a linear number of prices needs to be communicated.
- Linear prices, if perceived as a guideline, help bidders to easily find interesting items and bundles and allow for endogenous bidding (Kwon05).
- The perceived fairness of anonymous prices might be an issue in some applications.
- The number of auction rounds is much lower.

Overall, robustness of combinatorial auction formats against different bidding strategies is, aside from efficiency, fairness, and speed of convergence, an important topic auctioneers need to care about.

References

Ausubel, L., Milgrom, P.: Ascending auctions with package bidding. Frontiers of Theoretical Economics 1, 1–42 (2002)

Bichler, M., Shabalin, P., Pikovsky, A.: A computational analysis of linear-price iterative combinatorial auctions. Information Systems Research 20(1) (2009)

de Vries, S., Schummer, J., Vohra, R.: On ascending vickrey auctions for heterogeneous objects. Journal of Economic Theory 132, 95–118 (2007)

Kwon, R.H., Anandalingam, G., Ungar, L.H.: Iterative combinatorial auctions with bidder-determined combinations. Management Science 51(3), 407–418 (2005)

Mishra, D., Parkes, D.: Ascending price vickrey auctions for general valuations. Journal of Economic Theory 132, 335–366 (2007)

Parkes, D., Ungar, L.H.: Iterative combinatorial auctions: Theory and practice. In: 17th National Conference on Artificial Intelligence (AAAI 2000) (2000)

Porter, D., Rassenti, S., Roopnarine, A., Smith, V.: Combinatorial auction design. Proceedings of the National Academy of Sciences of the United States of America (PNAS) 100, 11153–11157 (2003)

Using Prediction Markets to Track Information Flows: Evidence from Google

Bo Cowgill[1], Justin Wolfers[2], and Eric Zitzewitz[3]

[1] Google, Inc, Mountain View, CA, USA
bcowgill@google.com
[2] Wharton School, University of Pennsylvania, Philadelphia, PA, USA
jwolfers@wharton.upenn.edu
[3] Dartmouth College, Hannover, NH, USA
eric.zitzewitz@dartmouth.edu

Abstract. Since 2005, Google has conducted the largest corporate experiment with prediction markets we are aware of. In this paper, we illustrate how markets can be used to study how an organization processes information. We show that market participants are not typical of Google's workforce, and that market participation and success is skewed towards Google's engineering and quantitatively oriented employees.

We document a number of biases in Google's markets, most notably an optimistic bias. Newly hired employees are on the optimistic side of these markets, and optimistic biases are significantly more pronounced on days when Google stock is appreciating. We also document a reverse favorite longshot bias, where rare events are underpriced by the market. Lastly, we find a bias against "extreme" events: In markets about a continuous variable, traders underprice the high and low ends of the spectrum and overprice the middle.

In the final section of our paper, we document correlated trading among employees who sit within a few feet of one another and employees with social or work relationships. These findings are interesting in light of recent research on the role of optimism in entrepreneurial firms, as well as recent work on the importance of geographic and social proximity in explaining information flows in firms and markets.

The current draft of the full paper can be read at http://bocowgill.com/GooglePredictionMarketPaper.pdf

Keywords: prediction markets, information transmission, bias, market efficiancy, peer effects.

S. Das et al. (Eds.): Amma 2009, LNICST 14, p. 3, 2009.
© ICST Institute for Computer Sciences, Social-Informatics and Telecommunications Engineering 2009

A Copula Function Approach to Infer Correlation in Prediction Markets

Agostino Capponi[1] and Umberto Cherubini[2]

[1] California Institute of Technology,
Division of Engineering and Applied Sciences
Pasadena CA 91125, USA
acapponi@caltech.edu.edu
[2] University of Bologna,
Department of Mathematical Economics
Bologna 40126, Italy
umberto.cherubini@unibo.it

Abstract. We propose the use of copula methods to recover the dependence structure between prediction markets. Copula methods are flexible tools to measure associations among probabilities because they encompass both linear and non linear relationship among variables. We apply the proposed methodology to three actual prediction markets, the Saddam Security, the market of oil spot prices and the Saddameter. We find that the Saddam Security is nearly independent of the oil market, while being highly correlated to the Saddameter. The results obtained appear to suggest that the Saddam Security prediction market may be noisy or overlooking some political factors which are instead considered by Saddameter and the oil market.

1 Introduction

Prediction markets, also called information markets, are a mechanism to aggregate information from widely dispersed economic actors with the objective to produce prediction about future events [7] . The market price reflects a stable consensus of a large number of opinions about the likelihood of the event. This has also been analytically verified in [9], where it is shown that for a large class of models, prediction market prices correspond with mean beliefs of market actors.

The range of applications is quite broad, from helping businesses making better investment decisions, to helping governments making better policy decisions on health care, monetary policy, etc... [1]. There are three main types of contracts:

- "winner-take-all": such contract is linked to the realization of a specific event, it costs some amount p to enter and pays off 1 if and only if that event occurs.
- "index": the payoff of such contract varies continuously based on a quantity that fluctuates over time, like the percentage of the vote received by a candidate.

S. Das et al. (Eds.): Amma 2009, LNICST 14, pp. 4–12, 2009.

 – "spread betting": traders bid on the cutoff that determine whether an event occurs, for example whether one football team will win by a number of points larger than a certain threshold.

The basic forms of these contracts will reveal the market consensus about the probability, mean or median of a specific event. However, by appropriate combinations of these markets, it is possible to evaluate additional statistics. For example, a family of winner-take-all contracts paying off if and only if a football team loses by $1, 2, \ldots, n$ points or wins by $1, 2 \ldots, n$ points would reveal the whole distribution on the outcome of the game.

The contracts described above depend on only one outcome. The same principles can be applied to contracts depending on the joint outcomes of multiple events. Such contracts are called contingent and provide insights into the correlation between different events. For instance, Wolfers and Zitzewitz [8] ran a market linked to whether George W. Bush would be re-elected, another market on whether Osama Bin Laden would be captured prior to the election, and a third market on whether both events would have occurred. Their findings were that there was a 91 % chance of Bush being reelected if Osama had been found, but a 67 % unconditional probability.

This paper proposes an approach to imply the correlation between two prediction markets, each linked to the realization of one specific event, without introducing a third market on the joint outcome of the two events. The rest of the paper is organized as follows. Section 2 recalls the basics of copula functions and proposes a methodology based on the Kendall function to recover the dependence structure of two prediction markets. Section 3 applies the methodology described in the previous section to infer the pairwise dependence structure of the prediction market of oil, the Saddam Security and the Saddameter. Section 4 concludes the paper.

2 Correlation Methods Based on Copulas

In this section we briefly recall the basics of copula functions and refer the reader to a textbook on copulas [5] and [2] for a more detailed treatment. Copula functions represent a general and flexible tool to measure association among probabilities. Association is a more general concept than correlation, encompassing both linear and non linear relationship among variables. In fact, while correlation measures linear relationships between variables, copula functions are invariant to whatever linear or non linear transformation of the variables. Applying linear correlation to the evaluation of non linear relationships may lead to outright blunders, as the following textbook example illustrates. Simply take x with standard normal distribution and defines $y = x^2$. Obviously x and y are associated, but computing covariance one gets

$$E(xy) - E(x)E(y) = E(x^3) = 0 \qquad (1)$$

Copula functions enable to overcome these problems by studying the dependence among marginal distributions instead of variables.

2.1   Copula Functions

The basic idea is based on the *principle of probability integral transformation*. Take variables X and Y with marginal distributions F_X and F_Y. The principle states that transformations $u \equiv F_X(X)$ and $v \equiv F_Y(Y)$ have uniform distribution in $[0, 1]$. Marginal distributions can be inverted, and if they are continuous such inverse is unique. Take now the joint distribution $H(X, Y)$. This can be written as

$$H(X, Y) = H(F_X^{-1}(u), F_Y^{-1}(v)) \equiv C(u, v) \qquad (2)$$

where $C(u, v)$ is called the copula function representing association between X and Y. So, every joint distribution can be written as a function taking the marginal distributions as arguments. On the contrary, one can prove that if $C(u, v)$ has suitable properties, then plugging univariate distributions into it generates a joint distribution. The requirements for $C(u, v)$ are summarized in what is known as Sklar's theorem, and are reported below.

Definition 1. *A copula function $C(u, v)$ has domain in the 2-dimension unit hypercube and range in the unit interval and satisfies the following requirements:*

1. **Groundedness:** $C(u, 0) = C(0, v) = 0$
2. **Identity of marginals:** $C(u, 1) = v$, $C(1, v) = u$
3. **2-increasing:** $C(u_1, v_1) - C(u_1, v_2) - C(u_2, v_1) + C(u_2, v_2) \geq 0$, *with* $u_1 > u_2, v_1 > v_2$.

In case when the random variables X and Y are perfectly positively associated, i.e. $Y = T(X)$, with T being a monotonically increasing transformation, we have that

$$C(u, v) = \min(u, v)$$

The value $\min(u, v)$ is also the maximum value that a copula can achieve and is commonly referred as *upper Fréchet bounds*. In case when the random variables X and Y are uncorrelated, we obtain the *independence copula* given by

$$C(u, v) = uv$$

In case when X and Y are perfectly negative associated, $Y = L(X)$, with L being a monotonically decreasing transformation then

$$C(u, v) = \max(u + v - 1, 0)$$

The value $\max(u + v - 1, 0)$ is commonly referred to as *lower Fréchet bounds*. Copula functions are non-parametric representations of the dependence structure and as such they are linked to non parametric association measures such as *Spearman's* ρ_S (also called rank correlation) or *Kendall's* τ (also called concordance measure). We next discuss the relation to the *Kendall's* τ coefficient which will be used in this paper. Let $(X_1, X_2), (Y_1, Y_2)$ be independent random pairs with distribution F and marginal distribution F_1 and F_2. Then the Kendall's τ is defined as

$$\tau = \mathbb{P}((X_1 - Y_1)(X_2 - Y_2) > 0) - \mathbb{P}((X_1 - Y_1)(X_2 - Y_2) < 0) \qquad (3)$$

The condition $(X_1 - Y_1)(X_2 - Y_2) > 0$ corresponds to (X_1, X_2), (Y_1, Y_2) being two concordant pairs, i.e. one of the two pairs has the larger value for both components, while the condition $(X_1 - Y_1)(X_2 - Y_2) < 0$ corresponds to (X_1, X_2), (Y_1, Y_2) being two discordant pairs in that for each pair one component is larger than the corresponding component of the other pair and the other component is smaller. Therefore, Kendall's τ is the difference between the probability of two random concordant pairs and the probability of two random discordant pairs. Eq. (3) can be developed further to illustrate the relation with the associated copula as follows

$$\tau = \Pr((X_1 - Y_1)(X_2 - Y_2) > 0) - \mathbb{P}((X_1 - Y_1)(X_2 - Y_2) < 0)$$
$$= 4 \int_0^1 \int_0^1 C(u_1, u_2) dC(u_1, u_2) - 1 \qquad (4)$$

This follows because

$$\begin{aligned} \mathbb{P}((X_1 - Y_1)(X_2 - Y_2) > 0) - \mathbb{P}((X_1 - Y_1)(X_2 - Y_2) < 0) &= \\ 2\mathbb{P}((X_1 - Y_1)(X_2 - Y_2) > 0) - 1 &= \\ 2(\mathbb{P}(Y_1 \leq X_1, Y_2 \leq X_2) + \mathbb{P}(X_1 \leq Y_1, X_2 \leq Y_2)) - 1 \end{aligned} \qquad (5)$$

which can be developed further as

$$\mathbb{P}(Y_1 \leq X_1, Y_2 \leq X_2) = \int_{\mathbb{R}} \int_{\mathbb{R}} \mathbb{P}(Y_1 \leq x_1, Y_2 \leq y_2) dC(F_1(x_1), F_2(x_2))$$
$$= \int_{\mathbb{R}} \int_{\mathbb{R}} C(F_1(x_1), F_2(x_2)) dC(F_1(x_1), F_2(x_2))$$
$$= \int_0^1 \int_0^1 C(u_1, u_2) dC(u_1, u_2)$$

The same argument applies to $\mathbb{P}(X_1 \leq Y_1, X_2 \leq Y_2)$ since (X_1, X_2) and (Y_1, Y_2) are two independent pairs from the same distribution, thus implying Eq. (4).

Let $(V_1, W_1), (V_2, W_2), \ldots, (V_n, W_n)$ available samples from the joint distribution H. Let us define the empirical measures

$$K_i^- = \sharp\{(X_j, Y_j) : X_j < X_i, Y_j < Y_i\}$$
$$K_i^+ = \sharp\{(X_j, Y_j) : X_j > X_i, Y_j > Y_i\}$$
$$K_i = K_i^- + K_i^+$$
$$\bar{K}_i = \sharp\{(X_j, Y_j) : X_j > X_i, Y_j < Y_i\} + \sharp\{(X_j, Y_j) : X_j < X_i, Y_j > Y_i\} \qquad (6)$$

where $\sharp$ is the cardinality of the set $\{.\}$. The population version of the association measure τ in Eq. (4) is given by

$$\tau_{pop} = \frac{\sum_{i=1}^n K_i - \bar{K}_i}{\sum_{i=1}^n K_i + \bar{K}_i} \qquad (7)$$

The class of copulas used in this paper to represent cross-section dependence is that of *Archimedean copulas*. Each copula of this class is generated using a function $\phi(x)$ as

$$C(u, v) \equiv \phi^{[-1]}(\phi(u) + \phi(v)) \qquad (8)$$

where $\phi(s)$ is called the *generator function* of the copula C. Here ϕ is a continuous, strictly decreasing function from $[0, 1]$ to $[0, \infty]$ such that $\phi(1) = 0$. The function $\phi^{[-1]}$ is called the pseudo-inverse of ϕ, it maps from the domain $[0, \infty]$ to $[0, 1]$ and is given by

$$\phi^{[-1]}(t) \begin{cases} \phi^{-1}(t) & 0 \leq t \leq \phi(0) \\ 0 & \phi(0) \leq t \leq \infty \end{cases} \tag{9}$$

2.2 Kendall Function

The principle of probability integral transformation is intrinsically univariate. Even if we know that $u = F_X(X)$ and $v = F_Y(Y)$ are uniformly distributed, we cannot say what is the joint distribution of $C(u, v)$. For the class of Archimedean copulas we may provide an interesting bivariate extension. For such class of copulas, the multivariate probability integral transformation is known as *Kendall function*, or *Kendall distribution*. The distribution of $C(u, v)$ is given by

$$KC(t) = \Pr(C(u, v) \leq t)$$
$$= t - \frac{\phi(t)}{\phi'^+(t)} \tag{10}$$

where $\phi'^+(t)$ denotes the right derivative of the copula generator.

We can compute the Kendall function for the copula $C(u, v) = \min(u, v)$ of perfect positive dependence as follows. We have

$$\begin{aligned} \Pr(\min(u, v) \leq t) &= \Pr(\min(F_X(X), F_Y(Y)) \leq t) \\ &= \Pr(F_X(X) \leq t) + \Pr(F_Y(Y) \leq t) - \Pr(F_X(X) \leq t, F_Y(Y) \leq t) \\ &= t + t - \Pr(\min(F_X(X), F_Y(Y)) \leq t) \end{aligned} \tag{11}$$

which implies that $\Pr(\min(u, v) \leq t) = t$, thus the Kendall function is given by

$$KC(t) = t \tag{12}$$

and consequently the distribution is uniform as in the univariate case. The generator $\phi(t) = -\log(t)$ generates the independence copula $C(u, v) = uv$, as it is next shown

$$\begin{aligned} C(u, v) &= \phi^{[-1]}(\phi(u) + \phi(v)) \\ &= e^{-(-\log(u) - \log(v))} \\ &= e^{\log(uv)} \\ &= uv \end{aligned} \tag{13}$$

and, using Eq. (10) we have that the Kendall function is given by

$$KC(t) = t - t \ln(t) \tag{14}$$

The generator $\phi(t) = 1 - t$ generates the copula $C(u, v) = \max(u + v - 1, 0)$ of perfect negative dependence, as it is next shown. It can be checked that $\phi^{-1}(t) = \max(1 - t, 0)$, which in turn implies

$$
\begin{aligned}
C(u, v) &= \phi^{[-1]}(\phi(u) + \phi(v)) \\
&= \max(1 - [(1 - u) + (1 - v)], 0) \\
&= \max(u + v - 1, 0)
\end{aligned} \tag{15}
$$

and, using Eq. (10) we have that the Kendall function is given by

$$
KC(t) = 1 \tag{16}
$$

Kendall functions can be used to evaluate the dependence functions, and to gauge how far or close they are to the case of perfect dependence or independence. This can be done by computing the sample equivalent of the Kendall function which is

$$
KC_i = \frac{K_i^-}{N - 1} \tag{17}
$$

Plotting the sample equivalent of the Kendall function can give a graphical idea of the dependence structure, and how it changes with levels of the marginal distributions. We will use this tool to calculate the correlation between three actual prediction markets in the next section.

3 An Empirical Analysis of Saddam Market and Oil

We apply the methodology described in earlier sections to infer correlation between the following three prediction markets:

- "Saddam Security". This was a contract offered on TradeSports paying $100 if Saddam Hussein were ousted from power by the end of June 2003.
- "Saddameter". This was an expert journalists estimate of the probability of the United States going to war with Iraq.
- Spot oil price. These are the spot prices of futures on oil.

For each pair of individual markets, the data are processed as follows. We first align the time series, so that they all start and end at the same date. Let n be length of the aligned time series. We construct an equally spaced grid of points in the interval $[0, 1]$ where the distance between two consecutive points is $\Delta = 0.05$. We then calculate a frequency distribution such that the frequency associated with the i-th point in the grid is the number of indices $j \in \{1, \ldots, n\}$ such that $KC_j \leq \Delta i$, where KC_j has been defined in Eq. (17). We refer to this frequency distribution as the empirical copula.

Figures 1, 2 and 3 display the empirical Kendall function KC_i relating respectively the Saddam security with the oil price, the Saddam security with the Saddameter and the Saddameter with the oil price.

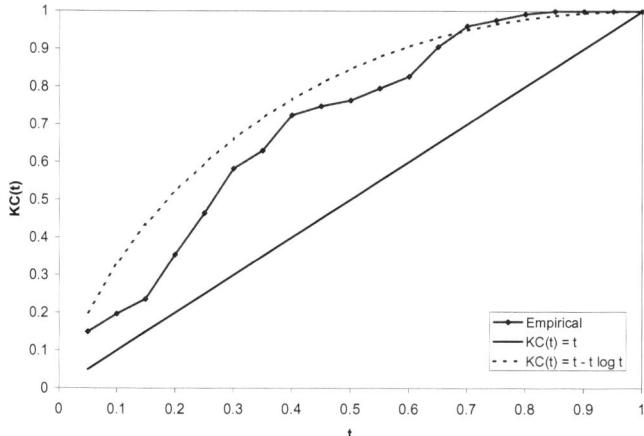

Fig. 1. Empirical kendall function $KC(t)$ between Saddam Security and oil price

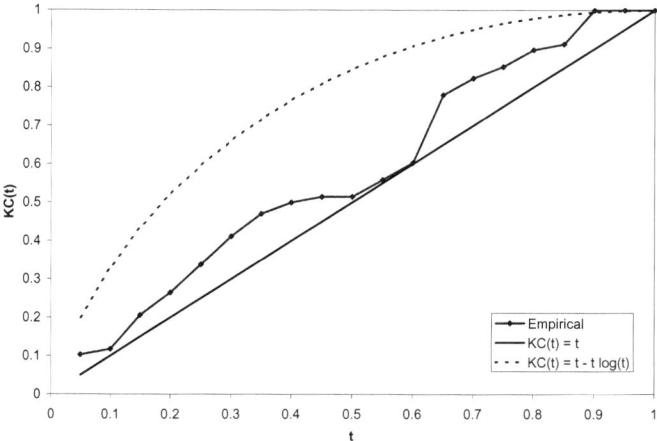

Fig. 2. Empirical kendall function $KC(t)$ between Saddam Security and Saddameter

It appears from Figure 1 that the prediction market evaluation of the Iraq war estimated by the Saddam security is nearly independent from the price of oil. This can be deduced by looking at the empirical copula which tends to lie closer to the curve $KC(t) = t - t\ln(t)$, representing the case of independence. The population value of the concordance measure τ_{pop} which evaluates to 25% reinforces the claim of independence. A closer look at Figure 1 shows that the independence is especially evidenced towards the tails (*tail independence*); when the Saddam security attributes large probability to Saddam being ousted, the oil market does not respond with increasing quotes of the oil price, and viceversa.

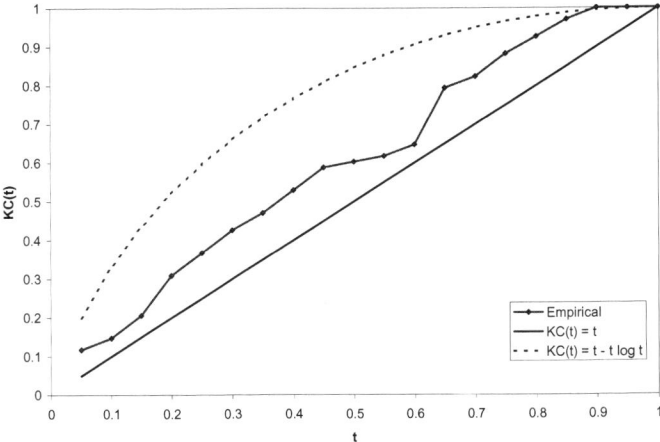

Fig. 3. Empirical kendall function $KC(t)$ between oil price and Saddameter

After repeating the same calculation for the case of the Saddam security and the Saddameter, we can see that an opposite argument takes place, and both expert evaluations and the Saddam security are in agreement with respect to the Iraq war. Figure 2, infact, shows that the empirical KC_i function tends to lie closer to the curve of perfect dependence $KC(t) = t$. For values of t closer to 0.5, the empirical curve lies on the curve of perfect dependence, whereas for extreme values of t, the empirical curve K_i tends to divert towards the curve of independence. The general behavior of positive correlation is also evidenced by the population value of the τ statistic which is given by $\tau_{pop} = 81\%$.

Figure 3 shows the empirical kendall function relating the Saddameter and the oil price. Even in this case, the empirical KC_i function tends to lie closer to the curve $KC(t) = t$, although not as close as for the case of the Saddam security and the Saddameter. Even in this case there is a tendency to move towards the curve of perfect independence for large values of t. The population value of the τ statistic which is given by $\tau_{pop} = 65\%$ confirms the graphical illustration of correlation.

The high correlation between Saddameter and oil price and the low correlation between the Saddam security and the oil price lead us to formulate the following hypotheses:

- *Hidden factors.* Journalist experts and the oil market may have looked at factors affecting the war in Iraq which were not considered by the prediction market of the Saddam security. In other words, there might have been a common political factor which drove both expert evaluations and traders in the oil market which was ignored by traders of the Saddam security.
- *Market Noise.* The prediction market of the Saddam security is noisy. Such noise makes the event of Iraq war independent of increases in the oil price. This noise may be caused by the presence of a sufficient number of "liquidity" traders who act like gamblers, and generate speculative bubbles by betting on unlikely

events in the hope of generating high returns. Those traders contrast with rational traders whose only motivation is expected return. If the market were only composed by the rational traders, then all information would be aggregated and fully reflected in prices and the No Trade Theorem would apply [4], [3].

4   Conclusions

In this paper, we have introduced an approach based on copula methods to infer dependence among prediction markets. Such methods are flexible enough to deal with non-linearity in the data. They are non-parametric, and thus do not require to impose any structure before performing the analysis. We have applied the methodology to infer dependence among three markets, the prediction market of the Saddam Security, the Saddameter, and the market of the spot future prices on oil. We found that the Saddam security is nearly independent of the oil market, while being highly correlated with the journalist expert evaluations provided by the Saddameter. We plan to perform a more detailed analysis on the correlation of those three markets as well as applying the proposed approach to other prediction markets and evaluate its robustness and predictive power.

Ackowledgments

Agostino Capponi would like to thank Prof. Justin Wolfers for sharing with him market quotes of the Saddam security, Saddameter, and spot oil prices.

References

1. Arrow, K.J., Forsythe, R., Gorham, M., Hahn, R., Hanson, R., Ledyard, J., Levmore, S., Litan, R., Milgrom, P., Nelson, F.D., Neumann, G.R., Ottaviani, M., Schelling, T.C., Shiller, R.J., Smith, V.L., Snowberg, E., Sunstein, C.R., Tetlock, P.C., Tetlock, P.E., Varian, H.R., Wolfers, J., Zitzewitz, E.: The promise of Prediction Markets. Science 16, 320(5878), 877–878 (2008)
2. Cherubini, U., Luciano, E., Vecchiato, W.: Copula Methods in Finance. Wiley Finance, Chichester (2004)
3. Grossman, S., Stiglitz, J.: Information and Competitive Price Systems. The American Economic Review 66(2), 246–253 (1976)
4. Milgrom, P., Stokey, N.: Information, trade and common knowledge. Journal of Economic Theory 26(1), 17–27 (1982)
5. Nelsen, R.: An Introduction to Copulas. Springer, Heidelberg (2006)
6. Wolfers, J., Zitzewitz, E.: Prediction Markets. Journal of Economic Perspectives 18(2) (2004)
7. Wolfers, J., Zitzewitz, E.: Experimental Political Betting Markets and the 2004 Election. The Economists' Voice 1(2) (2004)
8. Wolfers, J., Zitzewitz, E.: Information Markets: A New Way of Making Decisions in the Public and Private Sectors. In: Hahn, R., Tetlock, P. (eds.). AEI-Brookings Press
9. Wolfers, J., Zitzewitz, E.: Intrepreting Prediction Market Prices as Probabilities. Under Review in Review of Economics and Statistics

Manipulating Scrip Systems: Sybils and Collusion

Ian A. Kash[1], Eric J. Friedman[2], and Joseph Y. Halpern[1]

[1] Computer Science Dept., Cornell University
{kash,halpern}@cs.cornell.edu
[2] Sch. of Oper. Res. and Inf. Eng., Cornell University
ejf27@cornell.edu

Abstract. Game-theoretic analyses of distributed and peer-to-peer systems typically use the Nash equilibrium solution concept, but this explicitly excludes the possibility of strategic behavior involving more than one agent. We examine the effects of two types of strategic behavior involving more than one agent, sybils and collusion, in the context of scrip systems where agents provide each other with service in exchange for scrip. Sybils make an agent more likely to be chosen to provide service, which generally makes it harder for agents without sybils to earn money and decreases social welfare. Surprisingly, in certain circumstances it is possible for sybils to make all agents better off. While collusion is generally bad, in the context of scrip systems it actually tends to make all agents better off, not merely those who collude. These results also provide insight into the effects of allowing agents to advertise and loan money.

1 Introduction

Studies of filesharing networks have shown that more than half of participants share no files [1,11]. Creating a currency with which users can get paid for the service they provide gives users an incentive to contribute. Not surprisingly, scrip systems have often been proposed to prevent such free-riding, as well as to address resource-allocation problems more broadly. For example, KARMA used scrip to prevent free riding in P2P networks [17] and Mirage [4] and Egg [3] use scrip to allocate resources in a wireless sensor network testbed and a grid respectively.

Chun et al. [16] studied user behavior in a deployed scrip system and observed that users tried various (rational) manipulations of the auction mechanism used by the system. Their observations suggest that system designers will have to deal with game-theoretic concerns. Game-theoretic analyses of scrip systems (e.g., [2,8,10,14]) have focused on Nash equilibrium. However, because Nash equilibrium explicitly excludes strategic behavior involving more than one agent, it cannot deal with many of the concerns of systems designers. One obvious concern is collusion among sets of agents, but there are more subtle concerns. In a P2P network, it is typically easy for an agent to join the system under a number of different identities, and then mount what has been called a *sybil attack* [7]. While these concerns are by now well understood, their impact on a Nash equilibrium is not, although there has been some work on the effects of multiple identities in auctions [19]. In this paper we examine the effects of sybils and collusion on scrip systems. We show that if such strategic behavior is not taken into

S. Das et al. (Eds.): Amma 2009, LNICST 14, pp. 13–24, 2009.

account, the performance of the system can be significantly degraded; indeed the scrip system can fail in such a way that all agents even stop providing service entirely. Perhaps more surprisingly, there are circumstances where sybils and collusion can improve social welfare. Understanding the circumstances that lead to these different outcomes is essential to the design of stable and efficient scrip systems.

In scrip systems where each new user is given an initial amount of scrip, there is an obvious benefit to creating sybils. Even if this incentive is removed, sybils are still useful: they can be used to increase the likelihood that an agent will be asked to provide service, which makes it easier for him to earn money. This means that, in equilibrium, those agents who have sybils will tend to spend less time without money and those who do not will tend to spend more time without money, relative to the distribution of money if no one had sybils. This increases the utility of sybilling agents at the expense of non-sybilling agents. The overall effect is such that, if a large fraction of the agents have sybils (even if each has only a few), agents without sybils typically will do poorly. From the perspective of an agent considering creating sybils, the first few sybils can provide him with a significant benefit, but the benefits of additional sybils rapidly diminish. So if a designer can make sybilling moderately costly, the number of sybils actually created by rational agents will usually be relatively small.

If a small fraction of agents have sybils, the situation is more subtle. Agents with sybils still do better than those without, but the situation is not zero-sum. In particular, changes in the distribution of money can actually lead to a greater total number of opportunities to earn money. This, in turn, can result in an increase in social welfare: everyone is better off. However, exploiting this fact is generally not desirable. The same process that leads to an improvement in social welfare can also lead to a crash of the system, where all agents stop providing service. The system designer can achieve the same effects by increasing the average amount of money or biasing the volunteer selection process, so exploiting the possibility of sybils is generally not desirable.

Sybils create their effects by increasing the likelihood that an agent will be asked to provide service; our analysis of them does not depend on why this increase occurs. Thus, our analysis also applies to other factors that increase demand for an agent's services, such as advertising. In particular, our results suggest that there are tradeoffs involved in allowing advertising. For example, many systems allow agents to announce their connection speed and other similar factors. If this biases requests towards agents with high connection speeds even when agents with lower connection speeds are perfectly capable of satisfying a particular request, then agents with low connection speeds will have a significantly worse experience in the system. This also means that such agents will have a strong incentive to lie about their connection speed.

While collusion in generally a bad thing, in the context of scrip systems with fixed prices, it is almost entirely positive. Without collusion, if a user runs out of money he is unable to request service until he is able to earn some. However, a colluding group can pool there money so that all members can make a request whenever the group as a whole has some money. This increases welfare for the agents who collude because agents who have no money receive no service.

Furthermore, collusion tends to benefit the non-colluding agents as well. Since colluding agents work less often, it is easier for everyone to earn money, which ends up making everyone better off. However, as with sybils, collusion does have the potential of crashing the system if the average amount of money is high.

While a designer should generally encourage collusion, we would expect that in most systems there will be relatively little collusion and what collusion exists will involve small numbers of agents. After all, scrip systems exist to try and resolve resource-allocation problems where agents are competing with each other. If they could collude to optimally allocate resources within the group, they would not need a scrip system in the first place. However, many of the benefits of collusion come from agents being allowed to effectively have a negative amount of money (by borrowing from their collusive partners). These benefits could also be realized if agents are allowed to borrow money, so designing a loan mechanism could be an important improvement for a scrip system. Of course, implementing such a loan mechanism in a way that prevents abuse requires a careful design.

2   Model

Our model of a scrip system is essentially that of [14]. There are n agents in the system. One agent can request a service which another agent can volunteer to fulfill. When a service is performed by agent j for agent i, agent i derives some utility from having that service performed, while agent j loses some utility for performing it. The amount of utility gained by having a service performed and the amount lost by performing it may depend on the agent. We assume that agents have a *type* t drawn from some finite set T of types. We can describe the entire system using the tuple $(T, \boldsymbol{f}, n, m)$, where f_t is the fraction with type t, n is the total number of agents, and m is the average amount of money per agent. In this paper, we consider only *standard agents*, whose type we can characterize by a tuple $t = (\alpha_t, \beta_t, \gamma_t, \delta_t, \rho_t, \chi_t)$, where

- α_t reflects the cost of satisfying a request;
- β_t is the probability that the agent can satisfy a request
- γ_t measures the utility an agent gains for having a request satisfied;
- δ_t is the rate at which the agents discounts utility (so a unit of utility in k steps is worth only $\delta_t^{k/n}$ as much as a unit of utility now)—intuitively, δ_t is a measure of an agent's patience (the larger δ_t the more patient an agent is, since a unit of utility tomorrow is worth almost as much as a unit today);
- ρ_t represents the (relative) request rate (since not all agents make requests at the same rate)—intuitively, ρ_t characterizes an agent's need for service; and
- χ_t represents the relative likelihood of being chosen to satisfy a request. This might be because, for example, agents with better connection speeds are preferred.

The parameter χ_t did not appear in [14]; otherwise the definition of a type is identical to that of [14].

We model the system as running for an infinite number of rounds. In each round, an agent is picked with probability proportional to ρ_t to request service: a particular agent of type t is chosen with probability $\rho_t / \sum_{t'} f_{t'} n \rho_{t'}$. Receiving service costs some amount of scrip that we normalize to $1. If the chosen agent does not have enough scrip, nothing will happen in this round. Otherwise, each agent of type t is able to satisfy this request with probability β_t, independent of previous behavior. If at least one agent is able and willing to satisfy the request, and the requester has type t', then the requester gets a benefit of $\gamma_{t'}$ utils (the job is done) and one of the volunteers is chosen at random (weighted by the χ_t) to fulfill the request. If the chosen volunteer has type t, then that agent loses α_t utils, and receives a dollar as payment. The utility of all other agents is unchanged in that round. The total utility of an agent is the discounted sum of the agent's round utilities. To model the fact that requests will happen more frequently the more agents there are, we assume that the time between rounds is $1/n$. This captures the intuition that things are really happening in parallel and that adding more agents should not change an agent's request rate.

In our previous work [8,14], we proved a number of results about this model:

- There is an (ϵ-) Nash equilibrium where each agent chooses a threshold k and volunteers to work only when he has less than k dollars. For this equilibrium, an agent needs no knowledge about other agents, as long as he knows how often he will make a request and how often he will be chosen to work (both of which he can determine empirically).
- Social welfare is essentially proportional to the average number of agents who have money, which in turn is determined by the types of agents and the average amount of money the agents in the system have.
- Social welfare increases as the average amount of money increases, up to a certain point. Beyond that point, the system "crashes": the only equilibrium is the trivial equilibrium where all agents have threshold 0.

Our proofs of these results relied on the assumption that all types of agents shared common values of β, χ, and ρ. To model sybils and collusion, we need to remove this assumption. The purpose of the assumption was to make each agent equally likely to be chosen at each step, which allows entropy to be used to determine the likelihood of various outcomes, just as in statistical mechanics [12]. When the underlying distribution is no longer uniform, entropy is no longer sufficient to analyze the situation, but, as we show here, *relative entropy* [5] (which can be viewed as a generalization of entropy to allow non-uniform underlying distributions) can be used instead. The essential connection is that where entropy can be interpreted as a measure of the number of ways a distribution can be realized, relative entropy can be interpreted as a weighted measure where some outcomes are more likely to be seen than others. While this connection is well understood and has been used to derive similar results for independent random variables [6], we believe that our application of the technique to a situation where the underlying random variables are not independent is novel and perhaps of independent interest. Using relative entropy, all our previous results can be extended to the general case. The details of this extension are omitted for space, but are available in the full version [13].

This model does make a number of simplifying assumptions, but many can be relaxed without changing the fundamental behavior of the system (albeit at the cost of greater strategic and analytic complexity). For example, rather than all requests having the same value γ, the value of a request might be stochastic. This would mean that an agent may forgo a low-valued request if he is low on money. This fact may impact the threshold he chooses and introduces new decisions about which requests to make, but the overall behavior of the system will be essentially unchanged.

A more significant assumption is that prices are fixed. However, our results provide insight even if we relax this assumption. With variable prices, the behavior of the system depends on the value of β. For large β, where there are a significant number of volunteers to satisfy most requests, we expect the resulting competition to effectively produce a fixed price, so our analysis applies directly. For small β, where there are few volunteers for each request, variable prices can have a significant impact. In particular, sybils and collusion are more likely to result in inflation or deflation rather than a change in utility.

However, allowing prices to be set endogenously, by bidding, has a number of negative consequences. For one thing, it removes the ability of the system designer to optimize the system using monetary policy. In addition, for small β, it makes it possible for colluding agents to form a cartel to fix prices on a resource they control. It also greatly increases the strategic complexity of using the system: rather than choosing a single threshold, agents need an entire pricing scheme. Finally, the search costs and costs of executing a transaction are likely to be higher with variable prices. Thus, in many cases we believe that adopting a fixed price or a small set of fixed prices is a reasonable design decision.

3   Sybils

Unless identities in a system are tied to a real world identity (for example by a credit card), it is effectively impossible to prevent a single agent from having multiple identities [7]. Nevertheless, there are a number of techniques that can make it relatively costly for an agent to do so. For example, Credence uses cryptographic puzzles to impose a cost each time a new identity wishes to join the system [18]. Given that a designer can impose moderate costs to sybilling, how much more need she worry about the problem? In this section, we show that the gains from creating sybils when others do not diminish rapidly, so modest costs may well be sufficient to deter sybilling by typical users. However, sybilling is a self-reinforcing phenomenon. As the number of agents with sybils gets larger, the cost to being a non-sybilling agent increases and so the incentive to create sybils becomes stronger. Therefore measures to discourage or prevent sybilling should be taken early before this reinforcing trend can start. Finally, we examine the behavior of systems where only a small fraction of agents have sybils. We show that under these circumstances a wide variety of outcomes are possible (even when all agents are of a single type), ranging from a crash (where no service is provided) to an increase in social welfare. This analysis provides insight into the tradeoffs between efficiency and stability that occur when controlling the money supply of the system's economy.

When an agent of type t creates sybils, he does not change anything about himself but rather how other agents perceive him. Thus the only change to his type might be an increase in χ_t if sybils cause other agents to choose him more often. We assume that each sybil is as likely to be chosen as the original agent, so creating s sybils increases χ_t by $s\chi_t$. (Sybils may have other impacts on the system, such as increased search costs, but we expect these to be minor.) The amount of benefit he derives from this depends on two probabilities: p_e (the probability he will earn a dollar this round if he volunteers) and p_s (the probability he will be chosen to spend a dollar and there is another agent willing and able to satisfy his request). When $p_e < p_s$, the agent has more opportunities to spend money than to earn money, so he will regularly have requests go unsatisfied due to a lack of money. In this case, the fraction of requests he has satisfied is roughly p_e/p_s, so increasing p_e results in a roughly linear increase in utility. When p_e is close to p_s, the increase in satisfied requests is no longer linear, so the benefit of increasing p_e begins to diminish. Finally, when $p_e > p_s$, most of the agent's requests are being satisfied so the benefit from increasing p_e is very small. Figure 1 illustrates an agent's utility as p_e varies for $p_s = .0001$.[1] We formalize the relationship between p_e, p_s, and utility in the following theorem.

Theorem 1. *Let p_s be the probability that a particular agent is chosen to make a request in a given round and there is some other agent willing and able to satisfy it, p_e be the probability that the agent earns a dollar given that he volunteers, $r = p_e/p_s$, and k be the agent's strategy (i.e., threshold). In the limit as the number of rounds goes to infinity, the fraction of the agent's requests that have an agent willing and able to satisfy them that get satisfied is $(r - r^{k+1})/(1 - r^{k+1})$ if $r \neq 1$ and $k/(k+1)$ if $r = 1$.*

Proof. Since we consider only requests that have another agent willing and able to satisfy them, the request is satisfied whenever the agent has a non-zero amount of money. Since we have a fixed strategy and probabilities, consider the Markov chain whose states are the amount of money the agent has and the transitions describe the probability of the agent changing from one amount of money to another. This Markov chain satisfies the requirements to have a stationary distribution and it can be easily verified that the distribution gives the agent probability $r^i(1-r)/(1-r^{k+1})$ of having i dollars if $r \neq 1$ and probability $1/(k+1)$ if $r = 1$ [5]. This gives the probabilities given in the theorem.

Theorem 1 also gives insight into the equilibrium behavior with sybils. Clearly, if sybils have no cost, then creating as many as possible is a dominant strategy. However, in practice, we expect there is some modest overhead involved in creating and maintaining a sybil, and that a designer can take steps to increase this cost without unduly burdening agents. With such a cost, adding a sybil might be valuable if p_e is much less than p_s, and a net loss otherwise. This makes sybils a self-reinforcing phenomenon. When a large number of agents create sybils, agents with no sybils have their p_e significantly decreased. This makes them much worse off and makes sybils much more attractive to them. Figure 2 shows an example of this effect. This self-reinforcing quality means it is important to take steps to discourage the use of sybils before they become a problem. Luckily, Theorem 1 also suggests that a modest cost to create sybils will often be enough to prevent agents from creating them because with a well chosen value of m, few agents should have low values of p_e.

We have interpreted Figures 1 and 2 as being about changes in χ due to sybils, but the results hold regardless of what caused differences in χ. For example, agents may choose

---

[1] Except where otherwise noted, this and other figures assume that $m = 4$, $n = 10000$ and there is a single type of rational agent with $\alpha = .08$, $\beta = .01$, $\gamma = 1$, $\delta = .97$, $\rho = 1$, and $\chi = 1$. These values are chosen solely for illustration, and are representative of a broad range of parameter values. The figures are based on calculations of the equilibrium behavior. The algorithm used to find the equilibrium is described in [14].

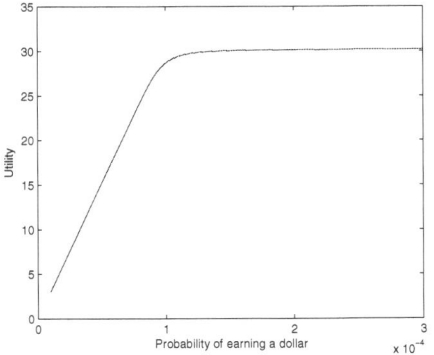

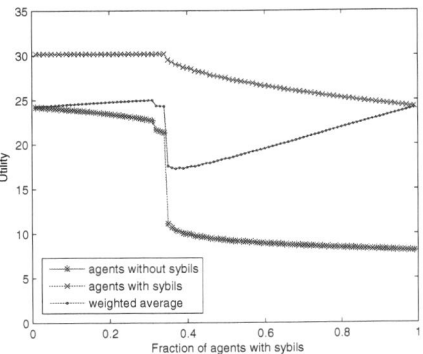

Fig. 1. The effect of p_e on utility   **Fig. 2.** The effect of sybils on utility

a volunteer based on characteristics such as connection speed or latency. If these characteristics are difficult to verify and do impact decisions, our results show that agents have a strong incentive to lie about them. This also suggests that the decision about what sort of information the system should enable agents to share involves tradeoffs. If advertising legitimately allows agents to find better service or more services they may be interested in, then advertising can increase social welfare. But if these characteristics impact decisions but have little impact on the actual service, then allowing agents to advertise them can lead to a situation like that in Figure 2, where some agents have a significantly worse experience.

We have seen that when a large fraction of agents have sybils, those agents without sybils tend to be starved of opportunities to work. However, as we saw in Figure 2, when a small fraction of agents have sybils this effect (and its corresponding cost) is small. Surprisingly, if there are few agents with sybils, an increase in the number of sybils these agents have can actually result in a *decrease* of their effect on the other agents. Because agents with sybils are more likely to be chosen to satisfy any particular request, they are able to use lower thresholds and reach those thresholds faster than they would without sybils, so fewer are competing to satisfy any given request. Furthermore, since agents with sybils can almost always pay to make a request, they can provide more opportunities for other agents to satisfy requests and earn money. Social welfare is essentially proportional to the number of satisfied requests (and is exactly proportional to it if everyone shares the same values of α and γ), so a small number of agents with a large number of sybils can improve social welfare, as Figure 3 shows. Note that this is not necessarily a Pareto improvement. For the choice of parameters in this example, social welfare increases when the agents create at least two sybils, but agents without sybils are worse off unless the agents with sybils create at least eight sybils. As the number of agents with sybils increases, they are forced to start competing with each other for opportunities to earn money and so are forced to adopt higher thresholds and this benefit disappears. This is what causes the discontinuity in Figure 2 when approximately a third of the agents have sybils.

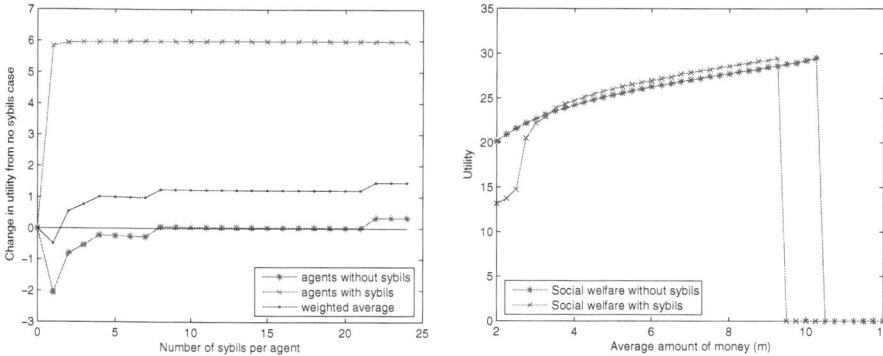

Fig. 3. Sybils can improve utility   **Fig. 4.** Sybils can cause a crash

This observation about the discontinuity also suggests another way to mitigate the negative effects of sybils: increase the amount of money in the system. This effect can be seen in Figure 4, where for $m = 2$ social welfare is very low with sybils but by $m = 4$ it is higher than it would be without sybils.

Unfortunately, increasing the average amount of money has its own problems. Recall from Section 2 that, if the average amount of money per agent is too high, the system will crash. It turns out that just a small number of agents creating sybils can have the same effect, as Figure 4 shows. With no sybils, the point at which social welfare stops increasing and the system crashes is between $m = 10.25$ and $m = 10.5$. If one fifth of the agents each create a single sybil, the system crashes if $m = 9.5$, a point where, without sybils, the social welfare was near optimal. Thus, if the system designer tries to induce optimal behavior without taking sybils into account, the system will crash. Moreover, because of the possibility of a crash, raising m to tolerate more sybils is effective only if m was already set conservatively.

This example shows that there is a significant tradeoff between efficiency and stability. Setting the money supply high can increase social welfare, but at the price of making the system less stable. Moreover, as the following theorem shows, whatever efficiencies can be achieved with sybils can be achieved without them, at least if there is only one type of agent. The theorem does require a technical condition similar in spirit to N-*replica economies* [15] to rule out transient equilibria that exist only for limited values of n. In our results, we are interested in systems $(T, \boldsymbol{f}, n, m)$ where T, $\boldsymbol{f}$, and m are fixed, but n varies. This leads to a small technical problem: there are values of n for which $\boldsymbol{f}$ cannot be the fraction of agents of each type nor can m be the average amount of money (since, for example, m must be a multiple of $1/n$). This technical concern can be remedied in a variety of ways; the approach we adopt is one used in the literature on N-replica economies.

Definition 1. A strategy profile $\boldsymbol{k}$ is an *asymptotic equilibrium* for a system $(T, \boldsymbol{f}, n, m)$ if for all n' such that $n' = cn$ for integer $c > 0$, $\boldsymbol{k}$ is a Nash equilibrium for $(T, \boldsymbol{f}, n', m)$.∎

Consider a system where all agents have the same type t. Suppose that some subset of the agents have created sybils, and all the agents in the subset have created the same number of sybils. We can model this by simply taking the agents in the subsets to have a new type s, which is identical to t except that the value of χ increases. Thus, we state our results in terms of systems with two types of agents, t and s.

Theorem 2. *Suppose that t and s are two types that agree except for the value of χ, and that $\chi_t < \chi_s$. If (k_t, k_s) is an asymptotic equilibrium for $(\{t, s\}, \boldsymbol{f}, m)$ with social welfare x, then there exists an m' and n' such that (k_s) is an asymptotic equilibrium for $(\{t\}, \{1\}, n', m')$ with social welfare at least x.*

Proof. We show this by finding an m' such that agents in the second system that play some strategy k get essentially the same utility that an agent with sybils would by playing that strategy in the first system. Since k_s was the optimal strategy for agents with sybils in the first system, it must be optimal in the second system as well. Since agents with sybils have greater utility than those without (they could always avoid volunteering some fraction of the time to effectively lower χ_s), social welfare will be at least as high in the second system as in the first.

Once strategies are fixed, an agent's utility depends only on $\alpha_t, \gamma_t, \delta_t, p_s$, and p_e. The first three are constants. Because the equilibrium is asymptotic, for sufficiently large cn, almost every request by an agent with money is satisfied (the expected number of agents wishing to volunteer is a constant fraction of cn). Therefore, p_s is essentially $1/cn$, his probability of being chosen to make a request. With fixed strategies, any value of p_e between 0 and β_t can be achieved by taking the appropriate m'. Take m' such that, if every agent in the second system plays k_s, the resulting p_e will be the same as the agents with sybils had in the original equilibrium. Note that p_s may decrease slightly because fewer agents will be willing to volunteer, but we can take $n' = cn$ for sufficiently large c to make this decrease arbitrarily small.

The analogous result for systems with more than one type of agent is not true. Consider the situation shown in Figure 2, where forty percent of the agents have two sybils. With this population, social welfare is lower than if no agents had sybils. However, the same population could be interpreted simply as having two different types, one of whom is naturally more likely to be chosen to satisfy a request. In this situation, if the agents less likely to be chosen created exactly two sybils each, the each agent would then be equally likely to be chosen and social welfare would increase. While changing m can change the relative quality of the two situations, a careful analysis of the proof of Theorem 2 shows that, when each population is compared using its optimal value of m, social welfare is greater with sybils. While situations like this show that it is theoretically possible for sybils to increase social welfare beyond what is possible by adjusting the average amount of money, this outcome seems unlikely in practice. It relies on agents creating just the right number of sybils. For situations where such a precise use of sybils would lead to a significant increase in social welfare, a designer could instead improve social welfare by biasing the algorithm agents use for selecting which volunteer will satisfy the request.

4   Collusion

Agents that collude gain two major benefits. The primary benefit is that they can share money, which simultaneously makes them less likely to run out of money and be unable to make a request and allows them to pursue a joint strategy for determining when to work. The secondary benefit, but important in particular for larger collusive groups, is that they can satisfy each other's requests. The effects of collusion on the rest of the system depend crucially on whether agents are able to volunteer to satisfy requests when they personally cannot satisfy the request but one of their colluding partners can. In a system where a request is for computation, it seems relatively straightforward for an agent to pass the computation to a partner to perform and then pass the answer back to the requester. On the other hand, if a request is a piece of a file it seems less plausible that an agent would accept a download from someone other than the person he expects and it seems wasteful to have the chosen volunteer download it for the sole purpose of immediately uploading it. If it is possible for colluders to pass off requests in this fashion, they are able to effectively act as sybils for each other, with all the consequences we discussed in Section 3. However, if agents can volunteer only for requests they can personally satisfy, the effects of collusion are almost entirely positive.

Since we have already discussed the consequences of sybils, we will assume that agents are able to volunteer only to satisfy requests that they personally can satisfy. Furthermore, we make the simplifying assumption that agents that collude are of the same type, because if agents of different types collude their strategic decisions become more complicated. For example, once the colluding group has accumulated a certain amount of money it may wish to have only members with small values of α volunteer to satisfy requests. Or when it is low on money it may wish to deny use of money to members with low values of γ. This results in strategies that involve sets of thresholds rather than a single threshold, and while nothing seems fundamentally different about the situation, it makes calculations significantly more difficult.

With these assumptions, we now examine how colluding agents will behave. Because colluding agents share money and types, it is irrelevant which members actually perform work and have money. All that matters is the total amount of money the group has. This means that when the group needs money, everyone in the group volunteers for a job. Otherwise no one does. Thus, the group essentially acts like a single agent, using a threshold which will be somewhat less than the sum of the thresholds that the individual agents would have used, because it is less likely that c agents will make ck requests in rapid succession than a single agent making k. Furthermore, some requests will not require scrip at all because they can potentially be satisfied by other members of the colluding group. When deciding whether the group should satisfy a member's request or ask for an outside volunteer to fulfill it, the group must decide whether it should pay a cost of α to avoid spending a dollar. Since not spending a dollar is effectively the same as earning a dollar, the decision is already optimized by the threshold strategy; the group should always attempt to satisfy a request internally unless it is in a temporary situation where the group is above its threshold.

Figure 5 shows an example of the effects of collusion on agents' utilities as the size of collusive groups increases. As this figure suggests, the effects typically go through three phases. Initially, the fraction of requests colluders satisfy for each other is small. This

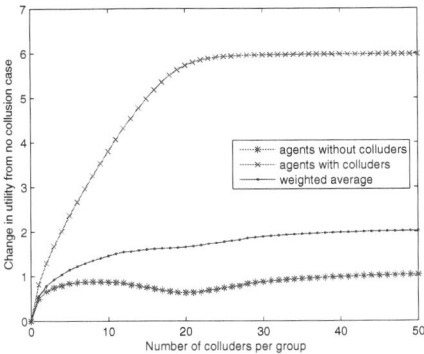

Fig. 5. The effect of collusion on utility

means that each collusive group must work for others to pay for almost every request its members make. However, since they share money, the colluders do not have to work as often as individuals would. Thus, other agents have more opportunity to work, and every agent's p_e increases, making all agents better off.

As the number of colluders increases, the fraction of requests they satisfy internally grows significant. We can think of p_s as decreasing in this case, and view these requests as being satisfied "outside" the scrip system because no scrip changes hands. This is good for colluders, but is bad for other agents whose p_e is lower since fewer requests are being made. Even in this range, non-colluding agents still tend to be better off than if there were no colluders because the overall competition for opportunities to work is still lower. Finally, once the collusive group is large enough, it will have a low p_s relative to p_e. This means the collusive group can use a very low threshold which again begins improving utility for all agents. Since collusion is difficult to maintain (the problem of incentivizing agents to contribute is the whole point of using scrip), we would expect the size of collusive groups seen in practice to be relatively small. Therefore, we expect that for most systems collusion will be Pareto improving. Note that, as with sybils, this decreased competition can also lead to a crash. However, if the system designer is monitoring the system and encouraging and expecting collusion she can reduce m appropriately and prevent a crash.

These results also suggest that creating the ability to take out loans (with an appropriate interest rate) is likely to be beneficial. Loans gain the benefits of reduced competition without the accompanying cost of fewer requests being made in the system. However, implementing a loan mechanism requires addressing a number of other incentive problems. For example, *whitewashing*, where agents take on a new identity (in this case to escape debts) needs to be prevented [9].

Acknowledgements. EF, IK, and JH are supported in part by NSF grant ITR-0325453. JH is also supported in part by NSF grant IIS-0812045 and by AFOSR grants FA9550-08-1-0438 and FA9550-05-1-0055. EF is also supported in part by NSF grant CDI-0835706.

References

1. Adar, E., Huberman, B.A.: Free riding on Gnutella. First Monday 5(10) (2000)
2. Aperjis, C., Johari, R.: A peer-to-peer system as an exchange economy. In: GameNets 2006: Proceeding from the 2006 Workshop on Game Theory for Communications and Networks, p. 10 (2006)
3. Brunelle, J., Hurst, P., Huth, J., Kang, L., Ng, C., Parkes, D., Seltzer, M., Shank, J., Youssef, S.: Egg: An extensible and economics-inspired open grid computing platform. In: Third Workshop on Grid Economics and Business Models (GECON), pp. 140–150 (2006)
4. Chun, B., Buonadonna, P., AuYung, A., Ng, C., Parkes, D., Schneidman, J., Snoeren, A., Vahdat, A.: Mirage: A microeconomic resource allocation system for sensornet testbeds. In: Second IEEE Workshop on Embedded Networked Sensors, pp. 19–28 (2005)
5. Cover, T., Thomas, J.: Elements of Information Theory. John Wiley & Sons, Inc., New York (1991)
6. Csiszár, I.: Sanov propery, generalized i-projection and a conditional limit theorem. The Annals of Probability 12(3), 768–793 (1984)
7. Douceur, J.R.: The sybil attack. In: Druschel, P., Kaashoek, M.F., Rowstron, A. (eds.) IPTPS 2002. LNCS, vol. 2429, pp. 251–260. Springer, Heidelberg (2002)
8. Friedman, E.J., Halpern, J.Y., Kash, I.A.: Efficiency and Nash equilibria in a scrip system for P2P networks. In: Proc. Seventh ACM Conference on Electronic Commerce (EC), pp. 140–149 (2006)
9. Friedman, E.J., Resnick, P.: The social cost of cheap pseudonyms. Journal of Economics and Management Strategy 10(2), 173–199 (2001)
10. Hens, T., Schenk-Hoppe, K.R., Vogt, B.: The great Capitol Hill baby sitting co-op: Anecdote or evidence for the optimum quantity of money? J. of Money, Credit and Banking 9(6), 1305–1333 (2007)
11. Hughes, D., Coulson, G., Walkerdine, J.: Free riding on Gnutella revisited: The bell tolls? IEEE Distributed Systems Online 6(6) (2005)
12. Jaynes, E.T.: Where do we stand on maximum entropy? In: Levine, R.D., Tribus, M. (eds.) The Maximum Entropy Formalism, pp. 15–118. MIT Press, Cambridge (1978)
13. Kash, I.A., Friedman, E.J., Halpern, J.Y.: Manipulating scrip systems: Sybils and collusion. arXiv:0903.2278v1
14. Kash, I.A., Friedman, E.J., Halpern, J.Y.: Optimizing scrip systems: Efficiency, crashes, hoarders and altruists. In: Proc. Eighth ACM Conference on Electronic Commerce (EC), pp. 305–315 (2007)
15. Mas-Colell, A., Whinston, M.D., Green, J.R.: Microeconomic Theory. Oxford University Press, Oxford (1995)
16. Ng, C., Buonadonna, P., Chun, B., Snoeren, A., Vahdat, A.: Addressing strategic behavior in a deployed microeconomic resource allocator. In: Third Workshop on Economics of Peer-to-Peer Systems (P2PECON), pp. 99–104 (2005)
17. Vishnumurthy, V., Chandrakumar, S., Sirer, E.G.: KARMA: a secure economic framework for peer-to-peer resource sharing. In: First Workshop on Economics of Peer-to-Peer Systems (P2PECON) (2003)
18. Walsh, K., Sirer, E.G.: Experience with an object reputation system for peer-to-peer file-sharing. In: Third Symp. on Network Systems Design & Implementation (NSDI), pp. 1–14 (2006)
19. Yokoo, M., Sakurai, Y., Matsubara, S.: The effect of false-name bids in combinatorial auctions: new fraud in internet auctions. Games and Economic Behavior 46(1), 174–188 (2004)

A Centralized Auction Mechanism for the Disability and Survivors Insurance in Chile

Gonzalo Reyes H.

Pensions Supervisory Authority of Chile
greyes@spensiones.cl

Abstract. As part of the pension reform recently approved in Chile, the government introduced a centralized auction mechanism to provide the Disability and Survivors (D&S) Insurance that covers recent contributors among the more than 8 million participants in the mandatory private pension system. This paper is intended as a case study presenting the main distortions found in the decentralized operation of the system that led to this reform and the challenges faced when designing a competitive auction mechanism to be implemented jointly by the Pension Fund Managers (AFP). In a typical bilateral contract the AFP retained much of the risk and the Insurance Company acted in practice as a reinsurer. The process to hire this contract was not competitive and colligated companies ended up providing the service. Several distortions affected competition in the market through incentives to cream-skim members by AFPs (since they bear most of the risk) or efforts to block disability claims. Since the price of this insurance is hidden in the fees charged by AFPs for the administration of individual accounts and pension funds there was lack of price transparency. Since new AFPs have no history of members' disability and mortality profile the insurance contract acted as a barrier to entry in the market of AFP services, especially when D&S insurance costs reached 50% of total costs. Cross-subsidies between members of the same AFP, inefficient risk pooling (due to pooling occurring at the AFP rather than at the system level) and regulatory arbitrage, since AFPs provided insurance not being regulated as an insurance company, were also present. A centralized auction mechanism solves these market failures, but also gives raise to new challenges, such as how to design a competitive auction that attracts participation and deters collusion. Design features that were incorporated in the regulation to tackle these issues, such as dividing coverage into predefined percentage blocks, are presented here.

S. Das et al. (Eds.): Amma 2009, LNICST 14, p. 25, 2009.
© ICST Institute for Computer Sciences, Social-Informatics and Telecommunications Engineering 2009

Information Feedback and Efficiency in Multiattribute Double Auctions

Kevin M. Lochner[1] and Michael P. Wellman[2]

[1] RentMineOnline.com 1690 Bay Street, Suite 302, San Francisco, CA 94123, USA
kevin@rentmineonline.com
[2] University of Michigan, Computer Science & Engineering, Ann Arbor, MI 48109-2121, USA
wellman@umich.edu

Abstract. We investigate tradeoffs among expressiveness, operational cost, and economic efficiency for a class of multiattribute double-auction markets. To enable polynomial-time clearing and information feedback operations, we restrict the bidding language to a form of multiattribute OR-of-XOR expressions. We then consider implications of this restriction in environments where bidders' preferences lie within a strictly larger class, that of complement-free valuations. Using valuations derived from a supply chain scenario, we show that an iterative bidding protocol can overcome the limitations of this language restriction. We further introduce a metric characterizing the degree to which valuations violate the substitutes condition, theoretically known to guarantee efficiency, and present experimental evidence that the actual efficiency loss is proportional to this metric.

1 Introduction

Multiattribute auctions mediate the trade of goods defined by a set of underlying features, or *attributes*. Bids express offers to buy or sell *configurations* defined by specific attribute vectors, and the auction process dynamically determines both the transaction prices and the configurations of the resulting trades. Most research on multiattribute auctions addresses the single-good procurement setting, in which a single buyer negotiates with a group of candidate suppliers [3,5,8,13]. Extending to the two-sided case offers the potential for enhanced efficiency, price dissemination, and trade liquidity.

In mediating the trade of multiple goods, it is often beneficial to consider preferences for bundles of goods. The exponentially sized offer specifications induced by such *combinatorial* valuations present difficulties, both for expression of agent valuations [15] and computation of optimal allocations [14]. For certain subclasses of multi-unit valuations, however, these problems may be tractable. Notably, for valuations satisfying the *gross substitutes* condition, it is well known that a price equilibrium exists, and such equilibria support efficient allocations. Furthermore, *market-based algorithms*—distributed iterative procedures that search over a space of linear prices—reliably converge to equilibrium under gross substitutes. Market-based algorithms that rely exclusively on offers for individual goods in effect provide a polynomial scheme for approximate computation of efficient allocations. In Section 3, we review gross substitutes and its relation to syntactically defined bidder valuation classes.

S. Das et al. (Eds.): Amma 2009, LNICST 14, pp. 26–39, 2009.

Since a configuration in multiattribute negotiation corresponds to a unique type of good, the class of multi-unit valuations for multiattribute goods is equivalent to the class of combinatorial valuations. The problem of multi-unit multiattribute allocation therefore inherits the hardness results derived for combinatorial auctions, but moreover applied to a cardinality of goods that is itself exponential in the number of attributes.

In Section 4, we present a two-sided multiattribute auction admitting polynomial-time clearing given a restricted bidding language. We extend a previously developed clearing algorithm [9] with a polynomial-time information feedback algorithm, enabling the implementation of market-based algorithms. Our mechanism thus extends some of the efficiency results of combinatorial auctions to multiattribute domains. We provide evidence that the inclusion of information feedback to our auction design successfully compensates for the lack of expressive power of our bidding language.

Theoretical work is largely silent on the efficiency of market-based algorithms given valuations violating gross substitutes. In Section 5, we present natural ways in which complement-free valuations may violate the gross substitutes condition, invalidating the efficiency guarantee of market-based approaches. In an effort to quantify the expected performance limits of our mechanism against a larger class of valuations, we introduce a new metric on bidder valuations, based on the severity by which valuations violate gross substitutes. We apply this metric to a family of valuations, derived from a supply chain manufacturing scenario, and present simulation results demonstrating a correlation between our metric and expected market efficiency.

2 Auction Preliminaries

Auctions mediate the trade of goods among a set of self-interested participants, or *agents*, as a function of agent messages, or *bids*. In a *multiattribute* auction, goods are defined by vectors of *attributes*, $a = (a_1, \ldots, a_m)$, $a_j \in A_j$. A *configuration*, $x \in X$, is a particular attribute vector. Each configuration can be thought of as a unique type of good. An *allocation*, $g \in G$, is a multiset of such goods, that is, a set possibly containing more than one instance of a given configuration.

Bids define one or more *offers* to buy or sell goods. An offer pairs an allocation and a *reserve price*, (g, p), where $g \in G$ and $p \in \Re^+$. For a buy offer, the reserve price indicates the maximum payment a buyer is willing to make in exchange for the set of goods comprising allocation g. Similarly, the reserve price of a sell offer defines the minimum payment a seller is willing to receive to provide allocation g.

A *bid*, $b \in B$, defines a set of offers (often implicitly) which collectively define an agent's reserve price over the space of allocations. We use the term *valuation* to designate any mapping from the space of allocations to the positive real numbers: $v : G \mapsto \Re^+$, hence a bid defines a valuation. For ease of explication, we use the function $r : G \times B \mapsto \Re^+$ to indicate the reserve price of a bid for a given allocation. The *bidding language* of an auction defines the space B of expressible bids.

Each bidder maintains a single active bid: b_i for buyer i and b_j for seller j. To *clear* the market, the auction computes a *global allocation* comprising an assignment of individual allocations and associated payments. The computed allocation must be

1. *feasible*: the set of goods allocated to buyers is contained in the set of goods supplied by sellers, and the net payments are nonnegative, and
2. *acceptable*: individual payments meet the reserve price constraints expressed in the bids of buyers and sellers.

We assume agents have *quasilinear* preferences over alternative allocation and payment outcomes. Buyer i has quasilinear utility function $u_i(g, p) = v_i(g) + p$, where valuation v_i defines the net change in buyer utility when supplied with a given allocation, and p denotes the net payments made to the buyer. Similarly, seller j has utility function $u_j(g, p) = -v_j(g) + p$, where valuation v_j is interpreted as a cost function for supplying allocations.

The allocations and payments determine the realized utilities of all agents. To the extent that bids accurately reflect valuations, an auction can use bids as proxies for underlying valuations, and maximize the objective function for the valuations expressed through bids. The extent to which bids do not accurately reflect agent valuations may induce inefficient (suboptimal) global allocations, as the maximization employs an inaccurate objective function. A bidding language which is syntactically unable to fully convey agent valuations may therefore impede efficiency. Since the complexity of optimizing global allocation increases with the expressiveness of the bidding language, we face a general tradeoff between computational and allocational efficiency.

In a *direct revelation* mechanism, each agent submits at most a single bid, in the form of a valuation, without receiving any information about the bids of other agents. In *iterative auctions*, agents revise their bids over time based on summary information provided by the auction about the current auction state. Summary information is typically derived from the clearing algorithm given the current auction state, informing agents of their current hypothetical allocations as well as *price quotes* indicating the minimum or maximum prices to buy or sell allocations [16].

3 Allocation with Complement-Free Valuations

We start by revisiting complexity results for combinatorial allocation, focusing on *complement-free* bidder valuations. Non-complementarity assumptions are commonly invoked in economic models, including diminishing marginal utilities for consumers and decreasing returns to scale for producers, and the substitutes condition for Walrasian equilibria [12]. The class of complement-free buyer valuations contains all valuations which are never *superadditive* over configurations.

Definition 1. *A buyer valuation is complement-free (CF) if for any g_a and g_b,*

$$v(g_a) + v(g_b) \geq v(g_a \cup g_b).$$

A seller valuation (cost function) is complement-free it is not subadditive over configurations, that is, the direction of the above inequality is reversed for sellers.

It is known that no polynomial clearing algorithm can guarantee better than a 2-approximation for the general class *CF* [7]. In the sequel, we present subclasses of *CF* of increasing complexity, borrowing both terminology and complexity results from Lehmann et al. [10], with notation amended slightly for multiattribute domains (where unique goods correspond to the configurations).

3.1 Syntactic Valuation Classes

Syntactic valuations are built from *atomic valuations* and operators on those valuations.

Definition 2. *The atomic valuation* (x, p) *gives the value p to any allocation containing a unit of configuration x, and value zero to all other allocations.*

Definition 3. *Let v_1 and v_2 be two valuations defined on the space G of allocations. The valuations $v_1 + v_2$ (OR) and $v_1 \oplus v_2$ (XOR) are defined by:*

$$(v_1 + v_2)(g) = \max_{g' \subseteq g}(v_1(g') + v_2(g \setminus g')),$$
$$(v_1 \oplus v_2)(g) = \max(v_1(g), v_2(g)).$$

Informally, the valuation $(v_1 + v_2)(g)$ divides up allocation g among valuations v_1 and v_2 such that the sum of the resulting valuations is maximized. The valuation $(v_1 \oplus v_2)(g)$ gives the entire allocation to v_1 or v_2, depending on which values g higher.

Subclasses of complement-free valuations are derived by placing restrictions on how the *OR* and *XOR* operators may be combined. Class *OS* valuations use only the *OR* operator over atomic valuations, thereby expressing additive valuations. Class *XS* valuations apply *XOR* over atomic valuations, thereby expressing substitute valuations. Any valuation composed of *OR* and *XOR* (applied in arbitrary order) falls into class *XOS*. The best approximation factor that can be guaranteed for *XOS* valuations in polynomial time is bounded above by 2, and below by $\frac{4}{3}$ [7].

3.2 *OXS* Valuations

Definition 4. *Applying ORs over XS valuations yields a valuation in class OXS.*

For example, as a buy bid, the valuation $(x_1, p_1) + [(x_2, p_2) \oplus (x_3, p_3)]$ expresses a willingness to buy x_1 at a price of p_1, and independently expresses a willingness to buy either x_2 at a price of p_2, or x_3 at a price of p_3 (but not both), giving the following acceptable allocations:

$$\{(x_1, p_1), (x_2, p_2), (x_3, p_3), (\{x_1, x_3\}, p_1 + p_3), (\{x_1, x_2\}, p_1 + p_2)\}.$$

If all bids express *OXS* valuations, the clearing problem can be formulated as a polynomial-time bipartite matching problem [9].

3.3 Gross Substitutes

To define valuations exhibiting gross substitutability, we must first introduce the concept of a demand correspondence. The following definitions are with respect to buyers.

Definition 5. *Given valuation v and configuration prices $\boldsymbol{p} = (p_{x_1}, \ldots, p_{x_n})$, demand correspondence $d(v \mid \boldsymbol{p})$ denotes the set of allocations that maximize $v(g) - \sum_{x \in g} p_x$.*

Definition 6. *A valuation v is of class GS if for any price vectors $\boldsymbol{p}$ and $\boldsymbol{q}$ with $p_i \le q_i \; \forall i$ and $g_1 \in d(v \mid \boldsymbol{p})$, there exists $g_2 \in d(v \mid \boldsymbol{q})$ such that $\{x \in g_1 \mid p_x = q_x\} \subset g_2$.*

Informally, *GS* requires that the demand for a given configuration be nondecreasing in the price of any other configuration. For sellers, the *supply* of a given configuration must be non*increasing* in the price of others.

Valuations satisfying the gross substitutes condition admit efficiency through *market-based algorithms*. Such algorithms operate by iteratively providing agents with price quotes, requiring that agents express *demand sets* reflecting their optimal consumption or production choices at the given prices. Demand sets are expressible in any bidding language of complexity equal to or greater than class *OS*. Prices are adjusted at each iteration based on the relative supply and demand of each type of good, until the market reaches equilibrium. Computationally, market-based algorithms provide a fully polynomial approximation scheme, with complexity that is polynomial in the number of bidders, goods, and the inverse of the approximation factor [10].

4   Call Market Implementation

In this section, we present the bidding language and algorithms supporting our multiattribute call market implementation. Though we focus here on the discrete configuration-based bidding language employed in our experimental study, both the clearing and information feedback algorithms admit more general bid forms [9,11].

4.1   Bidding Language

As discussed, multiattribute goods are defined in terms of possible *configurations* assigning values to attributes. The simplest multiattribute bidding unit expresses a maximum/minimum price at which to trade a given quantity of a single configuration.

Definition 7 (Multiattribute Point). *A multiattribute point of the form* (x, p, q) *indicates a willingness to buy up to total quantity q of configuration x at a unit price no greater than p (for $q > 0$). A negative quantity ($q < 0$) indicates a willingness to sell up to q units at a price no less than p.*

Participants in multiattribute auctions often wish to express flexibility over alternative configurations. For example, a computer buyer may be willing to accept various possibilities for processor type/speed, memory type/size/speed, etc., but at configuration-dependent reserve prices.

Definition 8 (Multiattribute *XR* Unit). *A multiattribute XR unit is a triple of the form* $((x_1, \ldots, x_N), (p_1, \ldots, p_N), q)$, *indicating a willingness to trade any combination of configurations* $(x_1, \ldots, x_N)$ *at unit prices* $(p_1, \ldots, p_N)$ *up to total quantity* $|q|$, *where* $q > 0$ *indicates a buy offer,* $q < 0$ *a sell offer.*

For example, given *XR* unit $((x_1, x_2, x_3), (p_1, p_2, p_3), 4)$, the allocation $\{x_1, x_1, x_2\}$ would be acceptable at total payment not greater than $p_1 + p_1 + p_2$.

In a slight abuse of notation, let $r(XR, x) = p$ select the unit reserve price for configuration x in the specified *XR* unit. Note that a multiattribute point is equivalent to an *XR* unit with single configurations and prices. To simplify our examples, we use the multiattribute point notation when an *XR* unit includes exactly one configuration.

Our final language construct is an *OR* extension of the *XR* unit.

Definition 9 (Multiattribute *OXR* Bid). *A multiattribute OXR bid, $\{XR_1, \ldots, XR_M\}$, indicates a willingness to trade any combination of configurations such that the aggregate allocation and payments to the bidder can be divided among the XR units such that each (g, p) pair is consistent with its respective XR unit.*

The bidding language constructs presented here can be classified within the syntactic framework presented above. A multiattribute point (x, p, q) expresses the valuation

$$\underbrace{(x, p) + (x, p) + \cdots}_{\text{total of } |q| \text{ atomic elements}}.$$

The additional quantity designation in a multiattribute point provides compactness over the equivalent *OR* expression when valuations are linear in quantity. A multiattribute *XR* unit with quantity q defines the valuation

$$\underbrace{[(x_1, p_1) \oplus \cdots \oplus (x_N, p_N)] + [(x_1, p_1) \oplus \cdots \oplus (x_N, p_N)] + \cdots}_{\text{total of } |q| \text{ XOR elements}}.$$

The multiattribute *XR* unit is less expressive than the general class *OXS* because it defines an *OR* over a set of identical *XOR* expressions, thus imposing a constraint that valuations be linear in quantity, and *configuration parity*, that is, the quantity offered by a bid is configuration-independent [9]. The *OXR* class is equivalent in expressiveness to *OXS*, though multiattribute *OXR* bids can be more compact and computationally convenient to the extent that valuations are linear in quantity.

4.2 Clearing

Previous work [9] explored the connection between bidding languages and clearing algorithms for this domain. Here we provide the main results but present them for only the *OXR* bidding language employed in the current study. The result holds for more general conditions on the bidding language as described in the earlier paper.[1]

Clearing the market requires finding the global allocation that maximizes the total trade surplus, which is the *Global Multiattribute Allocation Problem* (*GMAP*). For a certain class of bids, which includes *OXR* bids, *GMAP* can be divided into two discrete steps: identifying optimal bilateral trades (the *Multiattribute Matching Problem, MMP*), then maximizing total surplus as a function of those trades.

In the case of *OXR* bids, the multiattribute matching problem determines the optimal configuration x to trade between each pair of buy and sell *XR* units. For buy *XR* unit $XR_b = (configs^b, prices^b, q^b)$ and sell *XR* unit $XR_s = (configs^s, prices^s, q^s)$,

$$MMP_x(XR_b, XR_s) = \underset{x \in X}{\operatorname{argmax}}[r(XR_b, x) - r(XR_s, x)]. \tag{1}$$

The value achieved by the multiattribute matching solution (1) is called the *MMP surplus*, $MMP_s(XR_b, XR_s)$.

---

[1] The earlier paper [9] characterized bidding languages in terms of allocation constraints, rather than the complement-free hierarchy employed in the present work.

Define *BX* as the set of all *XR* units contained in the buyers' *OXR* bids, and *SX* the set of all *XR* units in the sellers' *OXR* bids. We start by solving *MMP* (1) for each pair in $BX \times SX$. *GMAP* is then formulated as a network flow problem, specifically the *transportation problem*, with source nodes *SX*, sink nodes *BX*, and link surplus (equivalently, negative link costs) equal to the values of MMP_s on $BX \times SX$. The optimal solution flow along a given link designates a quantity traded between the traders whose bids contain the respective *XR* units, and the configuration to be traded is the solution to MMP_x between the *XR* units.

4.3   Information Feedback

The decomposition of *GMAP* into *MMP* and subsequent network optimization can also be exploited for computing price quotes. To calculate quotes, we first find the required surplus (i.e., solution to MMP_s) for a new trade with a particular trader to be included in the efficient set. We can then determine the required price offer to that trader for any available configuration as a function of that required surplus. The computed price will be the quote for a $(configuration, trader)$ pair; taking the min/max over all sellers/buyers yields the ask/bid quote for a configuration.

This process is best described through example. Figure 1 depicts the *GMAP* formulation for a set of three sell offers (shown at left) and three buy offers (shown at right), all expressed as *XR* units. For example, XR_1 is an offer to sell a unit of either x_1 or x_2 at a price of 11. The solutions to MMP_s are indicated on the links connecting pairs of offers. The solution to *GMAP* is indicated by the bold links, in this case XR_5 and XR_2 trading one unit of x_2, and XR_3 and XR_6 trading one unit of x_2.

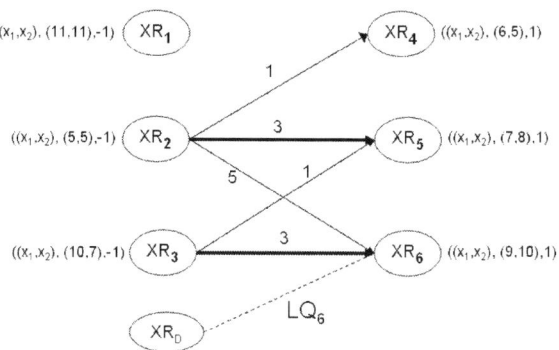

Fig. 1. *GMAP* formulation with three sell offers (left) and three buy offers (right). The optimal solution is indicated in bold. XR_D is a dummy node added for computing a link quote.

We now calculate the bid quote for x_2. As depicted in Figure 1, we first connect a dummy node (XR_D) to one of the existing buy nodes (node XR_6). We now calculate the minimum link surplus on the new edge that would increase the value of the optimal network flow. The computed link quote, LQ_6, is the trade surplus ($MMP_s(XR_6, XR_D)$) required for a new bid to trade with node XR_6. The link quote for each buy node must be calculated, producing a link quote for each $XR_k \in BX$.

The bid quote for a given configuration x is then:

$$\max_k(r(XR_k, x) - LQ_k).$$

Continuing with the example, Figure 2 depicts the computed link quotes. In this instance, the bid quote for configuration x_2 with XR_6 would be the offered price of 10, less the required link surplus of 4, producing a quote of 6 to transact *with that unit*. The bid quote for the configuration is the maximum over all the units, which is also 6. An offer price of 6 for x_2 would be sufficient to trade with either XR_6 or XR_5, as the quoted price for XR_5 would also be 6 (with a reserve price of 8 and required link surplus of 2).

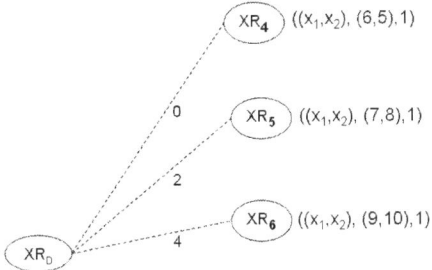

Fig. 2. Link quotes computed for a bid quote given the *GMAP* formulation of Figure 1

Finally, as confirmation that this process has produced a valid quote, we can consider the outcome of a new sell offer for a unit of x_2 at the quoted price. Figure 3 depicts this situation for the case that the new bid transacts with XR_5 (the algorithm will break the tie randomly) and shows that inclusion of the new bid has increased the trade surplus by 1 to a total of 7.

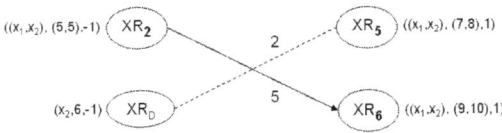

Fig. 3. *GMAP* solution for Figure 1 with a new sell offer at the quoted price

It is apparent from this formulation that once all link quotes have been determined, computing configuration quotes is proportional to the number of XR units. This implies that the complexity of a single configuration quote is invariant to the size of attribute space when the *GMAP-MMP* decomposition is applicable. Although computing quotes for all configurations entails complexity linear in the number of configurations, a bidder-driven query process for configuration quotes may still support market-based algorithm efficiency in large or continuous attribute domains.

Computing link quotes on the network flow graph is also achievable in polynomial time, using a specialization of the cycle-canceling algorithm [1]. Given that computation of a link quote requires perturbing the optimal network flow by quantity of only a single unit, the cycle-canceling algorithm can be adapted to a shortest-path algorithm, where an all-pairs shortest-path algorithm computes all required link quotes with complexity polynomial in the number of *XR* units. In practice, we require two iterations of the shortest-path algorithm, one iteration each for bid quotes and ask quotes.

5 Multiattribute Valuations

Our call market supports the direct expression of *OXS* valuations. However, many valuations natural for multiattribute domains fall outside of class *OXS*. For example, it is commonly desirable to ensure *homogeneity*, where all configurations in an allocation share values on one or more attributes [4]. Valuations placing higher values on homogeneous allocations are expressible with an *XOS* bidding language but not *OXS*.

The following example, inspired by a supply chain trading scenario [2], illustrates another situation where seller valuations fall outside of class *OXS*, and may violate *GS*.

Example 1. PCs are built from two components: cpu and memory. Assume that a manufacturer has one unit of $cpu = fast$, one unit of $cpu = slow$, one unit each for $memory \in \{large, medium, small\}$, with the following allowable configurations:

1. configuration x_1: $\{fast, large\}$
2. configuration x_2: $\{fast, medium\}$
3. configuration x_3: $\{slow, small\}$
4. configuration x_4: $\{slow, medium\}$

The production possibilities are then $\{x_1, x_4\}$, $\{x_1, x_3\}$, and $\{x_2, x_3\}$. The induced seller valuation is not expressible using an *OXS* language. The nearest *OXR* bid approximations require the seller to either overstate (bid B1) or understate (bids B2 and B3) his production capabilities:

$$(((x_1, x_2), (p_1, p_2), -1), ((x_3, x_4), (p_3, p_4), -1)) \tag{B1}$$

$$((x_1, p_1, -1), ((x_3, x_4), (p_3, p_4), -1)) \tag{B2}$$

$$(((x_1, x_2), (p_1, p_2), -1), (x_3, p_3, -1)) \tag{B3}$$

Assume that within the above production possibilities, the seller has a unit cost of 3 for all configurations, with total cost additive in unit cost. The exact *XOS* valuation would be

$$((x_1, 3) + (x_4, 3)) \oplus ((x_1, 3) + (x_3, 3)) \oplus ((x_2, 3) + (x_3, 3)).$$

This valuation is also not in class *GS*. Assume the prices of x_1 and x_4 are 5, and x_2 and x_3 are priced at 4. At these prices, the optimal production bundle is (x_1, x_4) which yields a surplus of 4. If the price of x_1 drops to zero, the optimal production bundle becomes (x_2, x_3), yielding a surplus of 2. Hence, the supply of x_4 decreases with a decrease in the price of x_1, which violates the gross substitutes condition for sellers.

6 A New Valuation Metric

Despite the limited expressive power of *OXS* bidding, we expect the iterative (market-based) version of our multiattribute auction to allocate effectively as long as valuations satisfy *GS*, or nearly do. To better characterize these situations, we introduce a measure of the *degree* to which a valuation violates the *GS* conditions.

6.1 Gross Substitutes Revisited

As defined above, *GS* requires that the demand for goods be nondecreasing in the prices of other goods. Intuitively, a price adjustment process will ultimately reach equilibrium if a price perturbation intended to reduce (increase) the demand of over(under)-demanded goods does not reduce (increase) the demand for other goods.

For valuation v satisfying *GS*, the demand correspondence condition holds for all price vectors and perturbations. Formally, given $d(v \mid p)$, the set of allocations maximizing $v(g) - \sum_{x \in g} p_x$, for all vectors of configuration prices $p = (p_{x_1}, \ldots, p_{x_n}) \in \Re_+^n$, and all single price perturbations $dp \in \Re_+^n$, for any $g_1 \in d(v \mid p)$ there exists $g_2 \in d(v \mid p + dp)$ such that $\{x \in g_1 \mid dp_x = 0\} \subset g_2$.

6.2 Gross Substitutes Violation

Let $p, dp \in \Re_+^n$, with $g_i \in d(v \mid p)$. The *gross substitutes violation* is given by:

$$GSV(v, p, dp, g_i) = \min_{g \in d(v \mid p + dp)} |\{x \in g_i \mid dp_x = 0\} \setminus \{x \in g \mid dp_x = 0\}|.$$

Intuitively, this measure counts the *number* of violations of the *GS* condition for a specific initial price vector and price change. Valuations satisfying *GS* have a violation count of zero for all initial prices, demand sets, and perturbations. Valuations that do not satisfy *GS* will have positive values of *GSV* for one or more combinations of (p, dp, g).

To simplify the exposition hereon, we assume $d(v \mid p)$ maps to a single g for any p, and use $x \in d(v \mid p)$ to indicate a good from that demand set. We next define the gross substitutes violation for a valuation and a price vector as the average *GSV* over all minimal single-price perturbations that ensure a new demand set.

$$GSV(v, p) = \frac{1}{n} \sum_{i=1}^{n} GSV(v, p, dp^i, d(v \mid p)),$$

where $dp^i = (0, \ldots, 0, dp_i, 0, \ldots, 0)$, and

$$dp_i = \min_{dp} dp \text{ s.t. } d(v \mid p) \neq d(v \mid (p_1, \ldots, p_i + dp, \ldots, p_n)).$$

Next, define the *expected* gross substitutes violation for a valuation as the expected value of *GSV* for random p ($p_i \sim U[0, \bar{p}]$),

$$EGSV(v) = E[GSV(v, p)].$$

The intuition behind using the expected *GSV* of a valuation (the average, rather than the maximum or minimum) is that any given run of a market-based algorithm traces a particular trajectory in price space, and the average violation is a proxy for the probability of seeing any specific violation.

7 Testing the *EGSV*-Efficiency Relationship

When *GS* holds, *EGSV* is zero, and market-based algorithms achieve full efficiency. Our hypothesis is that when the condition fails, realized efficiency will be decreasing in *EGSV*, all else equal. To evaluate this hypothesis, we employed a component-based model of configurations, as in Example 1. In this model, valuation complexity is determined by the *configuration structure*, as well as by the respective inventory levels and component costs of sellers.

For example, a valuation defined over configurations $\{x_1, x_2, x_3\}$ will violate *GS* to the extent that swapping production from one configuration to another requires additional components that are allocated to the third. Treating configurations $\{x_1, x_2, x_3\}$ as sets of components, assume that switching production from x_1 to x_2 requires additional components $x_2 \setminus x_1$. If an agent has no additional inventory of the components $(x_2 \setminus x_1) \cap x_3$ then the induced valuation will have a *GSV* of 1 for some price levels. In this way, variation both in the composition of configurations and the inventory levels of agents induces different levels of substitutability in agent valuations.

In the example above, if $x_2 \setminus x_1$ included two distinct components used by two different configurations, then the bidder valuation would have $GSV = 2$ for some price vectors, and thus a nonzero *EGSV* value. Conversely, if an agent had excess inventory of $x_2 \setminus x_1$, then the induced valuation would have $GSV = 0$ for all prices, and therefore the valuation would have an *EGSV* value of zero.

7.1 Valuation Generation

We generate a configuration structure by constructing random configurations until we have 20 distinct instances. For each configuration, we probabilistically include any one of eight unique components in the configuration (i.e., configurations may have variable numbers of components), while additionally requiring that any single configuration have at least three components. Given this structure, we randomly sample costs and inventory to generate a seller valuation. Seller inventories for each component are drawn independently from the discrete uniform distribution $[0, 3]$, while seller costs per component are drawn from the discrete uniform distribution $[30, 80]$.

We then evaluate *EGSV* for the induced valuation with respect to the price distribution from which agent valuations are drawn. For each price sample p,

1. determine the optimal production set $g^* = d(v \mid p)$,
2. identify all minimal single-price changes sufficient to change g^*, and
3. sum the *GS* valuations over these perturbations.

We iterate this process with random price samples until the standard error of *EGSV* is below .05. We generated a set of 100 valuations for each configuration structure, recording the costs and inventory, along with the *EGSV* value for each such valuation. We generated and evaluated seller valuations for 277 configuration structures, yielding a total of 27700 seller valuations.

7.2 Market Simulation

Each problem instance comprises 10 buyers and 10 sellers. For each configuration structure, we first sort the set of 100 generated seller valuations by *EGSV*. We define a unique problem instance for each contiguous set of 10 seller valuations, using the previously generated inventories and costs for each valuation, and taking the average *EGSV* value (denoted *aGSV*) of the 10 sellers to classify the problem instance. We thus generate 90 problem instances for each configuration structure.

We randomly generate buyer valuations for each problem instance. Each buyer has demand for two units, with full substitutability (i.e., will accept any combination of two goods at their reserve prices). Buyer reserve prices are drawn from the discrete uniform distribution $[400, 500]$ for each configuration.

For each problem instance, we first solve the allocation problem to determine the maximum achievable surplus. We then simulate bidding until quiescence, computing the fraction of efficiency achieved. To quantify the benefit of information feedback, we take the first iteration of bidding as the direct-revelation outcome. To evaluate the benefit of direct expression of substitutes—as in the *OXS* class supported by the *OXR* bidding language—we repeated the simulation with a class *OS* bidding language. Each problem instance thus produces four data points: one for each of $(direct, iterative) \times (OS, OXR)$.

Agents employ myopic best-response bidding, offering their true values at each iteration for a profit-maximizing set of goods. Given that the bidding language cannot fully express seller valuations, sellers are forced to approximate. To generate an *OS* bid, sellers find the feasible production bundle that maximizes profit at current prices (assuming a default price when quotes are not available). To generate an optimal *OXR* bid, sellers start with the optimal *OS* bid, and expand this to a feasible *OXR* bid.

7.3 Simulation Results

We aggregated the simulation results over all configuration structures and sorted the data by aGSV value into 10 bins. Figure 4 plots the average achieved fraction of

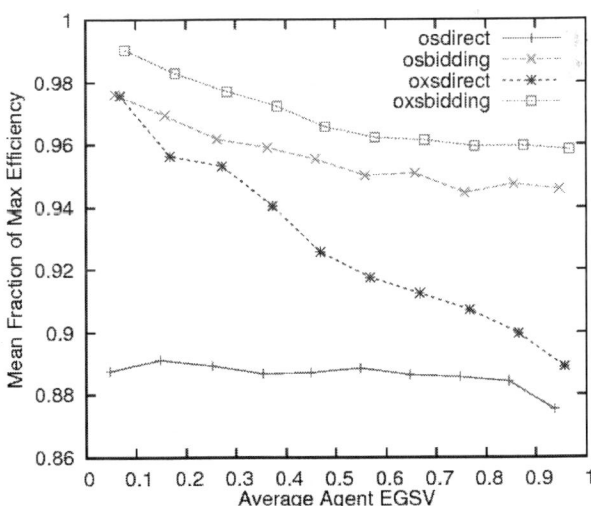

Fig. 4. Mean efficiency for average realized *EGSV*

maximal surplus as a function of aGSV value, for both direct-revelation and iterative mechanisms, for both the *OS* and *OXR* bidding languages.

For aGSV values close to zero, the substitutes condition is nearly satisfied for all valuations. Figure 4 confirms that iterative mechanisms perform well in this situation, averaging more than 97% efficiency for both *OXR* and *OS* bidding languages. The direct *OXR* mechanism (but not direct *OS*) also achieves this level of efficiency for low aGSV values. We conjecture that the majority of low *EGSV* valuations were also in class *OXS*, and therefore the ability to express substitutability through *OXR* bids is sufficient to achieve effective allocations without iteration.

Notable in Figure 4 is that the iterative mechanisms outperform the direct *OXR* mechanism by a margin that increases in aGSV value. We suspect this reflects valuations deviating further from class *OXS* with higher *EGSV* values. Despite increasing valuation complexity, the iterative mechanisms maintain a high level of efficiency, falling only to 95% as aGSV values reach 1. In this setting, information feedback is able to compensate for the lack of expressive power of a class *OXS* bidding language.

Finally, we observe that the iterative *OXR* mechanism outperforms the *OS* mechanism over all aGSV values. We hypothesize that the direct expression of substitutes allows the market-based algorithm to escape local maxima, as our mechanism does not implement a provably convergent market-based algorithm for *OS* bids.

8 Conclusions

We have introduced an implemented multiattribute call market with polynomial-time clearing and information feedback operations for a bidding language supporting a restricted class of combinatorial valuations. To our knowledge, this is the first call market of its kind presented in literature.

We analyzed the expected efficiency of our mechanism from the perspective of known hardness results derived for combinatorial auction settings, given complement-free bidder valuations. Using information feedback, iterative market-based algorithms can achieve efficient allocations given valuations satisfying the gross substitutes condition. Moreover, in some cases, iterative bidding can successfully compensate for expressive deficiencies imposed by a restricted bidding language.

Finally, we presented a new metric on bidder valuations, derived from the ways in which valuations violate *GS*. Experimental trials produce evidence that this metric correlates with the expected efficiency of market-based algorithms. The results suggest that measuring the degree of *GS* violation may provide a useful guide for predicting the performance of iterative bidding mechanisms, beyond the scope of environments for which theoretical guarantees apply.

References

1. Ahuja, R.K., Magnanti, T.L., Orlin, J.B.: Network Flows: Theory, Algorithms, and Applications. Prentice-Hall, Englewood Cliffs (1993)
2. Arunachalam, R., Sadeh, N.M.: The supply chain trading agent competition. Electronic Commerce Research and Applications 4, 63–81 (2005)

3. Bichler, M.: The Future of e-Markets: Multidimensional Market Mechanisms. Cambridge University Press, Cambridge (2001)
4. Bichler, M., Kalagnanam, J.: Configurable offers and winner determination in multi-attribute auctions. European Journal of Operational Research 160, 380–394 (2005)
5. Che, Y.-K.: Design competition through multidimensional auctions. RAND Journal of Economics 24, 668–680 (1993)
6. Cramton, P., Shoham, Y., Steinberg, R. (eds.): Combinatorial Auctions. MIT Press, Cambridge (2005)
7. Dobzinski, S., Nisan, N., Schapira, M.: Approximation algorithms for combinatorial auctions with complement-free bidders. In: Thirty-Seventh Annual ACM Symposium on Theory of Computing, pp. 610–618 (2005)
8. Engel, Y., Wellman, M.P.: Generalized value decomposition and structured multiattribute auctions. In: Eighth ACM Conference on Electronic Commerce, San Diego, pp. 227–236 (2007)
9. Engel, Y., Wellman, M.P., Lochner, K.M.: Bid expressiveness and clearing algorithms in multiattribute double auctions. In: Seventh ACM Conference on Electronic Commerce, pp. 110–119. Ann Arbor, MI (2006)
10. Lehmann, B., Lehmann, D., Nisan, N.: Combinatorial auctions with decreasing marginal utilities. Games and Economic Behavior 55, 270–296 (2006)
11. Lochner, K.M.: Multiattribute Call Markets. PhD thesis, University of Michigan (2008)
12. Mas-Colell, A., Whinston, M.D., Green, J.R.: Microeconomic Theory. Oxford University Press, New York (1995)
13. Parkes, D.C., Kalagnanam, J.: Models for iterative multiattribute procurement auctions. Management Science 51, 435–451 (2005)
14. Sandholm, T.: The winner determination problem. In: Cramton, et al [6]
15. Segal, I.: The communication requirements of combinatorial allocation problems. In: Cramton, et al [6]
16. Wurman, P.R., Wellman, M.P., Walsh, W.E.: A parametrization of the auction design space. Games and Economic Behavior 35, 304–338 (2001)

Impact of Misalignment of Trading Agent Strategy across Multiple Markets

Jung-woo Sohn, Sooyeon Lee, and Tracy Mullen

College of Information Sciences and Technology,
The Pennsylvania State University,
University Park, PA 16802-6823, USA
jwsohn@ist.psu.edu, sul131@psu.edu, tmullen@ist.psu.edu

Abstract. We examine the effect of a market pricing policy designed to attract high-valued traders in a multiple market context using JCAT software. Our experiments show that a simple change to pricing policy can create market performance effects that traditional adaptive trading agents are unable to recognize or capitalize on, but that market-policy-aware trading agents can generally obtain. This suggests as parameterized and tunable markets become more common, trading strategies will increasingly need to be conditional on each individual market's policies.

Keywords: market design, trading strategy, market selection strategy, multiple markets, pricing policy.

1 Introduction

The recent global financial crisis provides ample incentive for understanding market dynamics such as the spread of price volatility across multiple stock markets in different countries [2]. Yet, while real-world stocks can simultaneously be listed on the New York Stock Exchange (NYSE) and NASDAQ, classical economic theories and market microstructure typically assume a single market due to analytical complexity. The CAT market design competition [9] is aimed at encouraging more empirical investigation of such multiple market landscapes. Software market specialists compete against each other and are scored based on a combination of market share, profit from charging traders fees, and transaction rate. Each day software traders select a market to trade in, and then place shouts (i.e., bids or asks) in the market. Within this multiple market landscape, specialists must consider the intertwined effect of (1) market policies, (2) trader's market selection strategies and (3) trader's trading strategies.

In this paper, we show how a slight change in market microstructure aimed at attracting high-valued traders, whose bid is higher or ask is lower than the market equilibrium price, has unexpected implications both for trader's trading strategies and market selection strategies. In general, these high-valued traders are referred to as *intra-marginal traders*, while the low-value traders who bid lower or sell higher than the market equilibrium price are called *extra-marginal traders*. A market specialist who has numerous intra-marginal traders can more easily match bids and asks, resulting in a higher transaction rate score. Since intra-marginal traders are more likely to

S. Das et al. (Eds.): Amma 2009, LNICST 14, pp. 40–54, 2009.

be matched with another intra-marginal trader (depending on the matching algorithm), they can often achieve higher profit, and thus are more likely to return to the market. In addition, markets with more intra-marginal traders can charge higher fees without substantially reducing trader profit, thereby increasing the market profit per trader. However, some types of fees have the effect of driving away extra-marginal traders, thereby reducing the specialist's overall market share. For this initial work, we only consider free markets, and do not consider the impact of fees.

1.1 CAT Background

The CAT organizers provide a Java-based client-server platform, called JCAT [8] where software for specifying both trader and market specialist behavior is available. JCAT code can also be extended to incorporate new strategies and policies. A game parameter file determines numerous aspects of the game including how a trader's private value is set (randomly or fixed) or the number of trading days.

Traders are specified by selecting a trading strategy that governs traders' bidding behavior and a market-selection strategy that governs how traders choose between markets. Trading strategies in JCAT include truth-telling where each agent bids its private value, Zero-Intelligence-Plus (ZIP) [3], and Gjerstad-Dickhaut [5]. Examples of market selection strategies include a random strategy, where traders select a market to trade in randomly each day, and an adaptive learning strategy based on the N-armed bandit approach which selects the market with the best expected profit for $(1 - \epsilon)$ percent of the time, and randomly exploring other markets ϵ percent of the time.

Market specialists are specified by setting several market design parameters including the accepting, clearing, pricing, and charging policies. We focus here on the pricing policy, which determines how the transaction price is set when a bid and ask are matched.

1.2 Biased k-Pricing Policy

The PSUCAT team in the 2008 CAT competition adopted a biased k-pricing policy related to k-double auctions [11]. K-pricing policy sets the transaction price by dividing the bid-ask spread profit by the parameter k with values between 0 and 1. For example when k = 0.5, the transaction price is set halfway between the bid and ask. Our biased k-pricing aims to attract intra-marginal traders by giving more of the bid-ask spread profit to intra-marginal shouts matched with extra-marginal shouts. In this case, if k=0.9, 90% of the bid-ask spread profit goes to intra-marginal trader and only 10% to the extra-marginal trader. For example, suppose the market equilibrium price is 100, and a buyer with a bid of 130 (i.e., intra-marginal bid) is matched with a seller who asks 120 (i.e., extra-marginal ask), then the intra-marginal buyer gets 90% of the bid-ask spread profit and the transaction price is set to 121. An unbiased k-pricing market policy with k=0.5 would have set the transaction price of 125, giving buyer and seller equal profit. However, we found that intra-marginal traders did not effectively act on the biased k-pricing policy and thus did not favor our biased k-pricing market.

Further consideration showed that the biased k-pricing market introduces several inter-related issues. First, standard agent trading strategies are not attuned to market microstructure such as matching policy, pricing policy, and clearing policy. For example, the ZIP strategy assumes that the transaction price in the market is determined by the trading party who accepts the current market offer. When a buyer places a bid of $100 and a seller accepts it, the transaction price becomes $100. However, in the case of other market institutions such as sealed-bid auctions or when the market specialist pools and matches the shouts, the transaction price can differ. A seller who places an $80 ask might be matched with a buyer who placed a $100 bid, and the transaction price can be anywhere from $80 to $100.

Second, a related complication is that the adaptive trading strategies and the market selection strategies were not necessarily in alignment. For example, the ZIP trading strategy guides its adaptive behavior based on the profit margin, where profit is calculated as the difference between current shout price and private value. However, for market selection (other than random selection), traders calculate profit based on the difference between the transaction price and the trader's private value. So a buyer whose private value is 140, who bids 130, and who is matched with a seller for a transaction price of 121 will calculate its ZIP trading strategy profit as 140-130 or 10, while its market selection strategy will calculate the profit as 140 − 121 or 19. Thus adaptive traders cannot detect the extra profit achievable by the biased k-pricing policy and thus cannot use it to improve their trading results. On the other hand, intra-marginal truth-telling agents (who always bid their private value) are able to achieve high profits in the biased k-pricing market. Another inter-related problem is that the biased k-pricing market introduced a non-linear optimal bidding schedule for traders that simple adaptive behaviors did not handle well. We will discuss this aspect further in Section 3.

Lastly, even if the traders were able to discover the optimal bidding schedule, traditional trading agents do not distinguish their behavior based on what market they are in. Currently all trading strategies implicitly assume a single centralized market instead of conditioning their behavior on which market they are participating in.

In all of these cases, intelligent trading agents cannot fully exploit market microstructure to improve their profit. Since traders in the CAT tournament generally select markets based on the amount of the profit reaped so far, this can cause a trading agent to reach a suboptimal market selection decision. Thus our simple market policy change resulted in a more complex marketscape that simple adaptive behaviors were not able to optimize as well as a "stupid" truth telling trading strategy.

In this paper, we present a modeling approach that generalizes the inconsistency between the trading strategy and the market selection strategy. We also present a simulation result with a slight modification to the shout price variable in the original ZIP strategy, which we named ZIPK9Aware. The ZIPK9Aware agent was able to achieve higher overall profits by placing its shout price to take advantage of the market's biased k-pricing policy. We compare our ZIPK9Aware traders with standard ZIP and truth-telling traders.

2 Related Work

While continuous double auctions (CDA) are a well-established form of market institution, the complexity of even a single CDA means that human experiments and computer simulations are needed to more fully explore their properties. Smith [12] showed that even a few human traders in a continuous double auction would quickly converge to the equilibrium price. Gode and Sunder [6] introduced software zero-intelligence traders that randomly place shouts in a double auction market subject to a no-loss constraint. These zero-intelligence traders quickly converged to competitive equilibrium and suggested that equilibrium behavior can be achieved with extremely limited intelligence. However, Cliff [3] noticed that zero-intelligence traders owe their good trading behavior to regularities in the shape of the demand and supply curves, and thus did not generally perform well when the demand and the supply curve were asymmetric. Cliff designed his zero-intelligence plus (ZIP) traders on the assumption that traders should use at least some information about market conditions. ZIP traders collect information obtained from earlier shouts and trades and use it to adaptively set their target price margin for bidding. Similar to the ZIP strategy, the Roth-Erev (RE) strategy [10] is another adaptive trading strategy algorithm which adopts reinforcement learning. In the Gjestad-Dickhaut (GD) strategy [5], traders first collect the market history from an order book and use it to estimate the transaction success probability distribution, and then calculate the optimal shout price to maximize the expected profit.

All of the above work focused on traders trading a single good in a single market. If we extend this to trading a single good over multiple markets, some natural examples are financial markets and auction services. As Hasbrouck says [7] : "...Since a share of IBM is the same security whether purchased on the Midwest or Pacific Exchange, this is a particularly clear instance of multiple markets." Hasbrouck et al. focus on where price discovery occurs across multiple markets. In particular, they analyze NYSE and other regional exchanges seeking to establish "dominant" and "satellite" markets.

While the primary focus of this early research centered on price dynamics, Ellison et al. [4] view multiple market institutions as competing auctions. Their model analyzes what forces may cause markets to be concentrated. In particular, they consider why eBay, despite higher fees than its competitors, has achieved a dominant position in the online auction market while other auction services such as Yahoo have not acquired a large market share. Their preliminary results from two competing markets analysis imply that two markets will either co-exist or single market becomes dominant depending on various conditions.

Ellison et al. discuss the effects of attracting intra-marginal traders under various markets' charging policies. A typical strategy to acquire market share in auction services is to have zero listing fees. However, Ellison et al. found an unexpected effect of zero listing fees in Amazon or Yahoo was that they acquired numerous non-serious sellers with high reserve prices. In turn, serious buyers switched to other auction services. Ellision et al. also discuss the possibility that small markets can achieve high market efficiency by attracting high-value traders but do not proceed with further modeling analysis.

Westerhoff et al. [13] use an agent-based model to consider how small transaction taxes such as Keynes-Tobin tax can act to reduce market price volatility in a multiple market environment. They found that a transaction tax in one market reduces that market's price volatility but increases the volatility in the un-taxed market. However, when a transaction tax is imposed on both markets, then both markets show reduced price volatility. This suggests that market regulators in different markets could coordinate market transaction taxes to assist in controlling global market price volatility. Westerhoff's analysis also implies that once one market has imposed a transaction tax, other market regulators may wish to impose a transaction tax on their market so as not to face increased price volatility.

3 Problem Description

A trader in a multiple market scenario faces a two-stage process where the trader first selects the market in the first stage and places a shout in the market for the second stage. While the trading strategy and the market selection strategy are determined by the trading agent side, a market specialist selects one or more market policies such as pricing policy, in an effort to attract or keep certain types of traders based on their market selection strategy. However if a trader does not optimally trade in the given market, then a trader's market selection strategy will not necessarily operate optimally either. In this section, we characterize potential misalignments between agent trading strategies and market selection strategies due to the biased k-pricing policy.

3.1 Alignment of Trading Strategy and Market Selection Strategy

Trader profit is composed from two quantities. Let r be the transaction price determined by the market specialist, while s is the shout price and λ is the private value of the trading agent. The trading profit of $f(s)$ is composed from $|\lambda - s|$, the difference between the trader's private value and the shout price, and from $|s - r|$, the difference between the shout price and actual transaction price, so that $f(s) = |\lambda - s| + |s - r|$. Note that s is determined by the trader's bidding activity while r is determined by the market specialist.

Introducing the biased k-pricing policy to this framework, r becomes a function instead of a constant. Let s' be the price of the matched shout, then the profit now becomes $f(s) = |\lambda - s| + |s - r(s, s')|$. For the biased-k pricing, $r(s, s') = \{1 - k(s, s')\} s + k(s, s') s'$ where $k(s, s')$ becomes 0.9, 0.5, or 0.1 depending on whether s or s' are intra-marginal or extra-marginal shouts with respect to the current estimated equilibrium price p^*. Table 1 summarizes the k value assigned to the traders depending on their shout price s and s'. Note that traders can have some control over k by manipulating s in this framework.

However, the concept of trader's shout prices having an effect on market transaction price is not in alignment with most adaptive trading strategies. Traders consider k as exogenously determined by the market specialist. Thus, a trader's profit from an adaptive trading strategy actually stays as $f(s) = |\lambda - s| + |s - r|$, where r is a constant and thus $|s - r|$ cannot be easily maximized by varying s. ZIP traders even take this simplification one step further by ignoring any transaction-related profit so that their profit function becomes simply $f_{zip}(s) = |\lambda - s|$.

Table 1. k value assigned to each traders depending on the matched bid s and the ask s'

$(k(s,s')$ for bid, $1-k(s, s')$ for ask)	Intra-marginal ask $(s' < p^*)$	Extra-marginal ask $(s' > p^*)$
Intra-marginal bid $(s > p^*)$	$(k(s, s') = 0.5, 1-k(s, s') = 0.5)$	$(k(s, s') = 0.9, 1-k(s, s') = 0.1)$
Extra-marginal bid $(s < p^*)$	$(k(s, s') = 0.1, 1-k(s, s') = 0.9)$	trade not possible

Non-intelligent traders such as truth-tellers, who do not control their shout price s, implicitly get the best k. Since $s = \lambda$ for truth-tellers, their profit function $f(s)$ is reduced to $f_{TT}(\lambda) = |\lambda - r(\lambda, s')|$. For intra-marginal truth-tellers, k will be either 0.9 or 0.5 but never $k=0.1$. Now consider an imaginary price-taking trader for another example. Suppose the market provides a buyer with a quote of $p* + \varepsilon$ and provides a seller with the quote of $p* - \varepsilon$ in the hope that traders will place shouts honoring the biased k-pricing policy. The profit now becomes $f_{PT}(p*) = |\lambda - p*| + |s - r(p*, s')|$ where ε is infinitesimally small and can thus be ignored with k either 0.9 or 0.5 for intra-marginal traders. This exhibits an interesting implication that less intelligent trading strategies might actually optimize the profit coming from the transaction price set by the market in some cases.

3.2 Modifications in Trading Strategy under Biased K-Pricing Policy

Now we consider how an adaptive trading strategy could successfully act on the biased k-pricing policy. Figure 1 shows the profit schedule of an intra-marginal buyer when faced with the biased k-pricing policy. For simplicity, the seller is assumed to be an intra-marginal truth-teller. Note that the profit drop occurs around the market equilibrium price $p*$ when the intra-marginal buyer is matched with another intra-marginal seller.

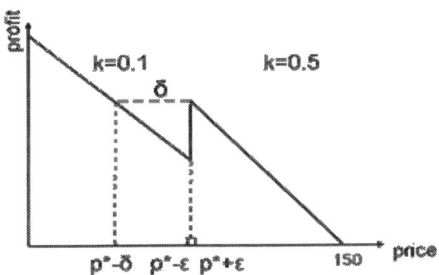

Fig. 1. Profit drop caused by biased K9 pricing policy for a buyer bidding at price (x-axis) with private value 150. (Seller is an intra-marginal truth-teller with the private value 50)

The buyer's profit drop from placing an intra-marginal bid at $p* + \varepsilon$ to placing an extra-marginal bid at $p* - \varepsilon$ is $f(p*+ \varepsilon) - f(p* - \varepsilon) =(0.9 - 0.5)(p* - s')= 0.4 (p* - s')$ where ε is infinitesimally small. (Since $p*$ itself is on the line between the intra-marginal and extra-marginals, we use ε to force the shout to be either intra-marginal or extra-marginal.) Now the profit drop depends on the matching shout price of s' since $p*$ can be assumed to be constant. Next consider the profit drop, δ in Figure 1, where the bidder is penalized for placing a bid in the range $[p* - \delta, p*]$. To avoid being penalized, the trader must estimate δ. By equating $f(p*) = \lambda - (0.5 p* + 0.5 s')$ with $f(p* - \delta) = \lambda - (0.9 (p - \delta) + 0.1 s')$, we have $\delta = (4/9)(p* - s')$. Since s' cannot be directly calculated by the trader, its trading strategy has to estimate s'. Figure 2 shows a 3-d plot of an intra-marginal buyer's profit across various possible combinations of (s, s'). As can be seen from the graph, profit estimation can become complex, which in turn results in the difficulty of estimation of δ.

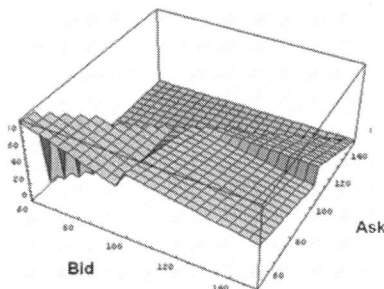

Fig. 2. 3-dimensional plot of the trade profit for intra-marginal buyer with the private value of 150 under K9 pricing policy. Market equilibrium price p* = 100.

One simple way to tackle this situation is to assume a constant average δ for trading agents regardless of their private value and to check its effect with simulation experiments. In the next section, we show that adjusting the trading strategy so that traders do not place shouts in the interval of $[p^* - \delta, p^*]$ for buyers and $[p^*, p^* + \delta]$ for sellers can lead to a significant increase in trader profit and improve the decisions of the trader's market selection strategy.

4 Experimental Setup

For the experimental setup we used the Java-based JCAT market simulation platform. The trading population consisted of 50 buyers and 50 sellers. We compare the trading strategies truth-telling, ZIP, and our ZIPK9Aware strategy. The ZIPK9Aware strategy is a simple modification to the original ZIP strategy that avoids placing a shout in the range of $[p^* - \delta, p^*]$ for buyers and $[p^*, p^* + \delta]$ for sellers. For each shout generated in that range, the ZIPK9Aware strategy recasts it as a bid for p^* to avoid facing a profit drop.

Private values for buyers and sellers are randomly drawn from a uniform distribution of [50, 150], giving a theoretical market equilibrium price of 100. Each trader is endowed with 5 goods to trade per game day. A single game lasts for 100 days and each game day has 10 rounds. To ensure our results are consistent, each game result is averaged over 10 trials.

We used a market selection strategy based on N-armed bandit problem in which the trader selects the market with the highest expected profit for 90% of the time and randomly selects markets 10% of the time. The market accepting policy is the same as the New York Stock Exchange (NYSE) spread-improvement rule, which is typically found in CDA experiments, and requires that new bids (or asks) must be higher (or lower) than the current best bid (or ask) in market. The market clears matching bid and ask whenever it finds the best matching pair.

4.1 ZIP Strategy and ZIPK9Aware Strategy

To make trading strategies detect the profit drop caused by our biased k-pricing policy, we modified the original ZIP strategy into our ZIPK9Aware strategy. See Bagnal and Toft [1] for further algorithmic details on ZIP strategy.

One issue related to the experimental design with ZIPK9Aware strategy is that a ZIPK9Aware trader needs to have an appropriate estimate of the profit drop δ in advance to determine the threshold where it changes the bidding strategy from ZIP to ZIPK9Aware and vice versa. Under the assumption that $p^*=100$ and the matching shout price of s' is drawn from a uniform distribution of [50, 100], the expected value of δ becomes 100/9 since $E[\delta] = (4/9)(p^* - E[s'])= 4/9 (100 - 75)$. However, we chose 15 which is slightly greater than 100/9 so that ZIPK9Aware traders are more likely to behave as ZIPK9Aware traders in uncertain situations, not as the original ZIP traders. Thus a ZIPK9Aware buyer will update its shout to $p^* + \varepsilon$ when its shout price is within $[p^* - \delta, p^*]$ interval to avoid being penalized by $k = 0.1$ pricing when it crosses the K5/K9 threshold. Similarly, a ZIPK9Aware seller will lower its shout price to $p^* - \varepsilon$ for the shout price interval of $[p^*, p^* + \delta]$. As discussed in the previous section, finding the optimal δ can be complicated so we initially took a simplification approach by using constant δ to check if our general idea is feasible.

5 Experimental Results

5.1 Truth-Teller Case

We start our experiments using truth-telling traders as a baseline case. Figure 4 shows the CAT tournament score for the averaged result of 10 experimental runs where M1 is a free market with K5 pricing policy and PSUCAT is a free market with the biased K9 pricing policy.[1] In CAT tournaments, each market's scores are composed of market share, profit ratio earned by the market, and the transaction success rate for the shouts placed in the market. These three factors are weighted equally and added together to evaluate the market's performance for a game day. Since neither market charges fees, their profit score is zero. The total score becomes the sum of each market's market share and transaction success rate scores. Figure 4 shows that the K9 market acquires both a larger market share and a higher transaction ratio than the K5 market. This implies that more truth-tellers prefer the K9 market, which in turn implies that traders earn more profit in the K9 market. However, large number of traders do not necessarily result in better market performance because markets cannot match trades when there are a high percentage of extra-marginal traders as pointed out by Niu et al. [9]. Figure 5 shows buyer's average private values for each market, while Figure 6 shows seller's average private values.

---

[1] For convenience, we introduce the term K5 market for the market with the biased k-pricing policy of $k=0.5$ and K9 market for the market with the biased k-pricing policy set to $k=0.9$. The data shown in the Figures are all averaged on 10 experimental runs.

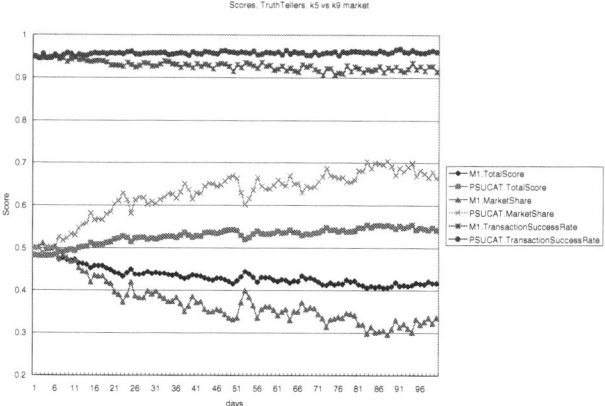

Fig. 4. JCAT scores for K5 (M1) and K9 (PSUCAT) markets with truth-telling traders

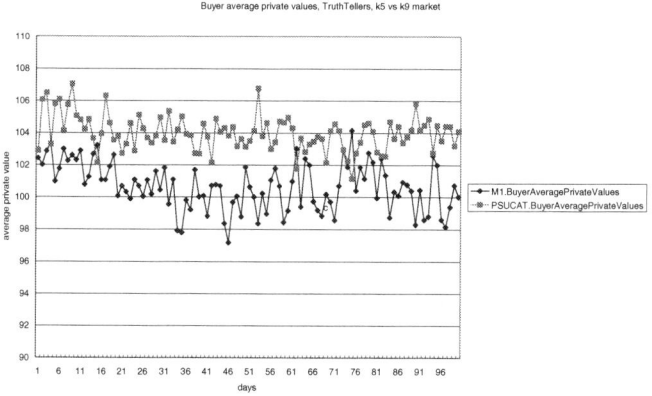

Fig. 5. Average private values for K5 (M1) and K9 (PSUCAT) markets with truth-telling buyers

As shown above in Figure 5 and 6, the K9 market attracts intra-marginal traders who have higher private values for buyers and lower private values for sellers than the theoretical private value average of 100. The separation effect looks more evident from the seller side in figure 6.

5.2 ZIP Trader Case

Figure 7 shows the game scores for standard ZIP traders, who do not recognize the biased k-pricing policy. Unlike the previous case, the market share between K5 market and K9 market does not show any significant difference nor does the transaction rate score. Clearly the K9 pricing policy did not make a significant difference in market selection behavior for ZIP traders.

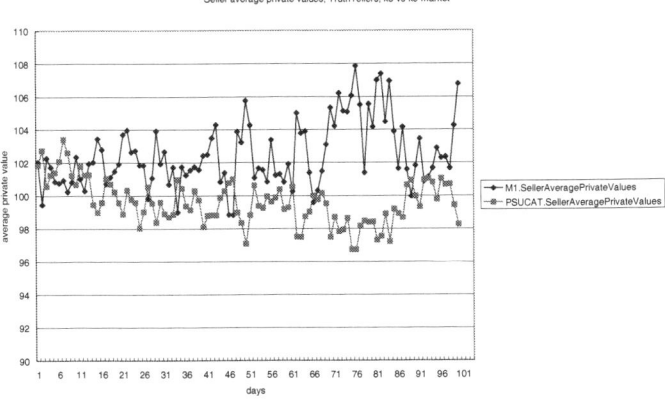

Fig. 6. Average private values for K5 (M1) and K9 (PSUCAT) markets with truth-telling sellers

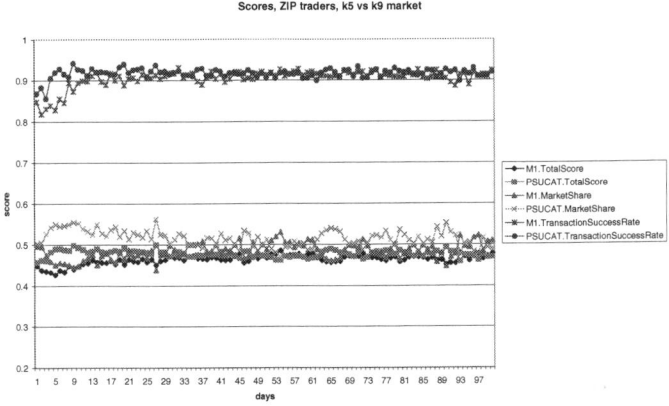

Fig. 7. JCAT scores for K5 (M1) and K9 (PSUCAT) markets with ZIP traders

Average trader private value plots shown in Figures 8 and 9 do not demonstrate a significant difference either. While the K9 market seems to attract intra-marginal traders until day 20, the separation effect is weak after that. This same pattern can be observed for both buyers and sellers.

5.3 ZIPK9Aware Case

Figure 10 shows the CAT game scores using ZIPK9Aware traders. The K9 market acquires traders from the K5 market, which can be seen from increasing market share and the transaction success rate drop for K5 market. Investigation of the log file for individual trial runs revealed that the K5 market temporarily experienced zero transactions in several runs which drove down the averaged value.

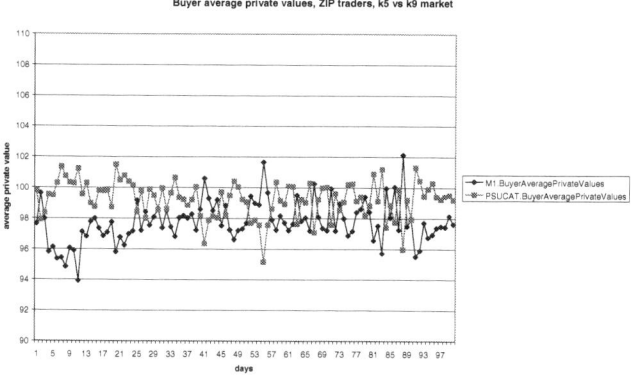

Fig. 8. Average private values for K5 (M1) and K9 (PSUCAT) markets with ZIP buyers

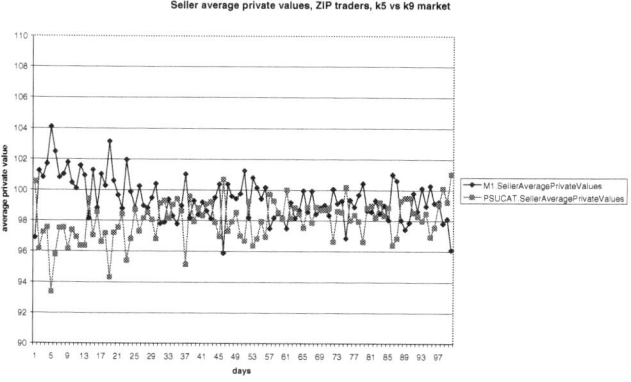

Fig. 9. Average private values for K5 (M1) and K9 (PSUCAT) markets with ZIP sellers

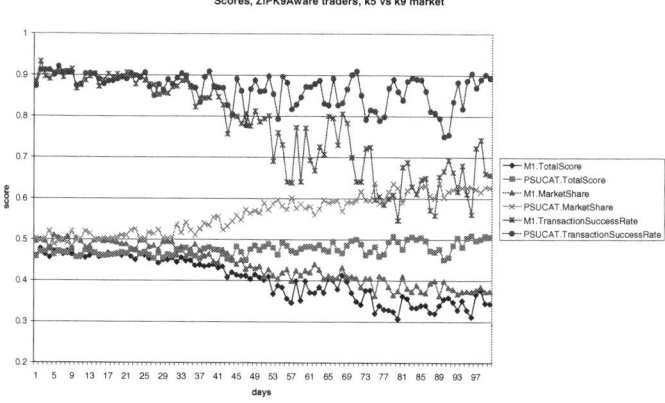

Fig. 10. JCAT scores for K5 (M1) and K9 (PSUCAT) markets with ZIPK9Aware traders

Figure 11 shows a different pattern than the previous cases where average private values for K5 market buyers show a steep drop compared to the increase in average for K9 market buyers. This implies that the K9 market not only attracts intra-marginal traders but also extra-marginal traders as well. Given the sudden transaction success rate drop in K5 market, apparently the K9 market successfully attracted both intra-marginal and extra-marginal traders by providing minimum trading profit for extra-marginal traders and not harming trading profits for intra-marginal traders. The result is that both intra-marginal and extra-marginal traders are attracted to the K9 market since extra-marginal traders can sometimes make trades with intra-marginals when attracted to the K9 market.

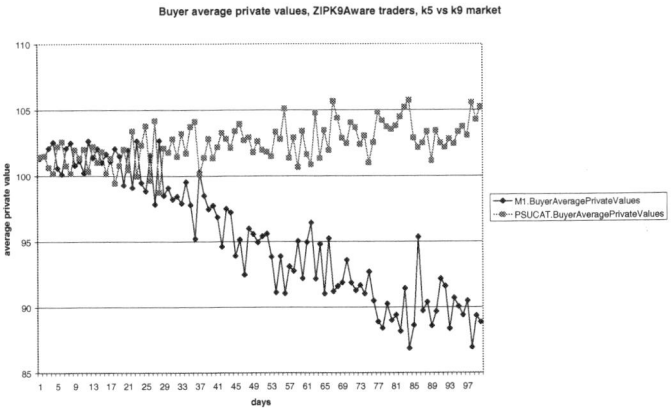

Fig. 11. Average private values for K5 (M1) and K9 (PSUCAT) markets with ZIPK9Aware buyers

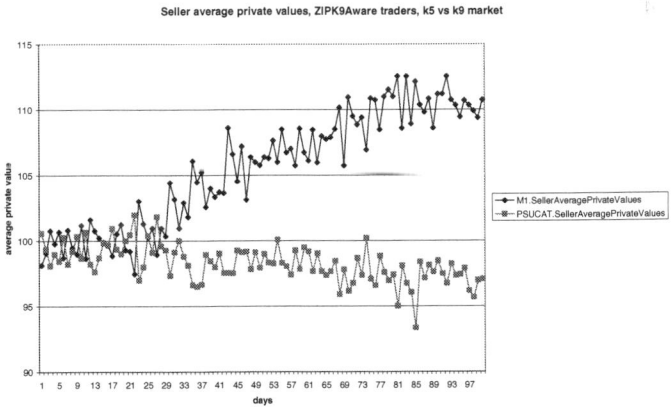

Fig. 12. Average private values for K5 (M1) and K9 (PSUCAT) markets with ZIPK9Aware sellers

However, in examining the 10 trials individually, 2 trials produced the opposite result where the main market share went to the K5 market and intra-marginal traders were attracted to the K5 market. While we do not understand exactly why these anomalous cases occur, one possibility is that we applied the same δ of 15 for all traders. The optimal size of δ depends on the matching shout, but we simplified the estimation process for the purpose of our experimental trials. In future work, we plan to increase the sophistication of the ZIPK9Aware trading agents to see if this produces the desired effect.

5.4 Comparison of Total Profit

Table 2 shows the comparison of average total profit reaped in K5 and K9 markets. ZIPK9Aware traders are making the highest total profits in K9 market. While ZIP traders still seem to make more profits on average in K9 market, the profit increase about 38,513 is relatively small compared to truth-teller and ZIPK9Aware trader cases of 172,193 and 132,993 respectively.

Table 2. Average total profit (and the stdev) earned in K5 market vs. K9 market over 10 runs. Numbers preceded by * denote that comparison is significant at 95% confidence interval under the null hypothesis that average total profit in K9 market is the same to that of K5 market. # denote significance at 90% confidence interval.

	K5 market			K9 market		
	Buyer	Seller	Total	Buyer	Seller	Total
Truth-teller	72105 (11157)	72105 (11157)	144207	*162715 (22014)	*153684 (24384)	316400
ZIP	108736 (35416)	110771 (22531)	219507	137581 (25907)	120439 (20127)	258020
ZIPK9Aware	108108 (80470)	99682 (66016)	207790	#159721 (83000)	*181061 (76755)	340783

In the K9 market, more profit is made by the agent trading strategies of truth-telling and ZIPK9Aware than the original ZIP traders. This implies that trading policies in alignment with the K9 pricing policy actually allows traders to acquire larger total profit.

5 Conclusion and Future Work

Our preliminary research demonstrates that market policy and agent trading behavior need to be aligned to perform effectively. We explore the implications of a biased k-pricing policy with $k=0.9$, called K9 pricing policy. This policy was aimed at incentivizing intra-marginal traders to favor the K9 market. We showed that the ZIP trading strategy is not in alignment with the K9 pricing policy since it docs not consider the actual transaction price, which in turn leads to sub-optimal bidding decisions and market selection decisions. We developed a ZIPK9Aware trading strategy, a simple modification on ZIP strategy to verify our argument that trading strategies should be in alignment with market policies. With ZIPK9Aware traders, the K9 market was able to attract more intra-marginal traders than the K5 market. The K9 market was also able to attract more market share and total profit. Thus our experiments show that a simple change to market pricing policy can create market performance effects that traditional adaptive trading agents are unable to recognize or capitalize on, but that market-policy-aware trading agents can obtain (most of the time). This suggests as more parameterized and tunable markets become more common, trading strategies will increasingly need to be conditional on a specific market's policies.

Alternatively from the market design point of view, our experimental results also suggest the not-surprising idea that market specialists should consider participant's trading strategies when selecting a market policy. In our experiment, truth-tellers and ZIPK9Aware traders were able to take advantage of the biased k-pricing policy, while ZIP traders were not. An interesting question is whether human traders would recognize the impact of a k=0.9 pricing policy and act more like ZIPK9Aware traders or if they would act more like ZIP traders. A more general question is how well humans (or trading agents) can recognize and act on various market policies to maximize profits.

Finally, our results suggest that traditional adaptive trading strategies should be extended to keep a separate trading history for each market. This may be especially important if markets have substantially different market policies.

For future work, we plan to test our result with other intelligent trading strategies such as GD and RE. In addition, we will investigate further anomalous cases where the ZIPK9Aware trading agents prefer the K5 market over the K9 market. While our preliminary work used the same δ values for all ZIPK9Aware traders, we will optimize δ for each individual trader's private value. We also plan to investigate our results using human subjects to see if they can recognize the K9 profit drop around the market equilibrium price.

Acknowledgements

We would like to thank the CAT tournament organizers and developers for stimulating our research work as well as our reviewers for their helpful comments.

References

1. Bagnall, A.J., Toft, I.E.: Zero Intelligence Plus and Gjestad-Dickhaut Agents for Sealed Bid Auctions. In: Workshop on Trading Agent Design and Analysis, part of International Conference on Autonomous Agents and Multiagent Systems (AAMAS 2004), pp. 59–64 (2004)
2. Buchanan, M.: This Economy Does Not Compute, The New York Times (October 1, 2008)
3. Cliff, D.: Minimal-intelligence Agents for Bargaining Behaviors in Market-based Environments, Technical Report, HP-97-91, Hewlett-Packard Research Laboratories, Bristol, England (1997)
4. Ellison, G., Fudenberg, D., Mobius, M.: Competing Auctions. Journal of the European Economic Association 2(1), 30–66 (2004)

5. Gjestad, S., Dickhaut, J.: Price formation in Double Auctions. Games and Economic Behavior 22, 1–29 (1998)
6. Gode, D.K., Sunder, S.: Allocative Efficiency of Markets with Zero-Intelligence Traders: Market as partial substitute for Individual Rationality. The Journal of Political Economy 101(1), 119–137 (1993)
7. Hasbrouck, J.: One Security, many Markets: Determining the Contributions to Price Discovery. Journal of Finance 50(4), 1175 (1995)
8. http://jcat.sourceforge.net
9. Niu, J., McBurney, P., Gerding, E., Parsons, S.: Characterizing Effective Auction Mechanisms: Insights from the 2007 TAC Market Design Competition. In: Proceedings of the 7th International Conference on Autonomous Agents and Multi-Agent Systems, Estoril, Portugal (2008)
10. Roth, A., Erev, I.: Learning in Extensive-form Games: Experimental Data and Simple Dynamic Models in the Intermediate Term. Games and Economic Behavior 8(1), 164–212 (1995)
11. Satterthwaite, M.A., Williams, S.R.: Bilateral Trade with the Sealed Bid K-Double Auction: Existence and Efficiency. Journal of Economic Theory 48, 107–133 (1989)
12. Smith, V.L.: An Experimental Study of Competitive Market Behavior. Journal of Political Economy 70(2), 111–137 (1962)
13. Westerhoff, F.H., Dieci, R.: The Effectiveness of Keynes-Tobin Transaction Taxes when Heterogeneous Agents can Trade in Different Markets: A Behavioral Finance Approach. Journal of Economic Dynamics and Control 30(2), 293–322 (2006)

Market Design for a P2P Backup System
(Extended Abstract)⋆

Sven Seuken[1], Denis Charles[2], Max Chickering[2], and Sidd Puri[2]

[1] School of Engineering & Applied Sciences, Harvard University,
Cambridge, MA 02138
seuken@eecs.harvard.edu
[2] Microsoft Live Labs, Redmond, WA 98004
{cdx,dmax,siddpuri}@microsoft.com

Keywords: market design, P2P, backup systems, resource exchange markets.

1 Introduction: A P2P Backup System

Peer-to-peer (P2P) backup systems are an attractive alternative to server-based systems because the immense costs of large data centers can be saved by using idle resources on millions of private computers instead. This paper presents the design and theoretical analysis of a market for a P2P backup system. While our long-term goal is an open resource exchange market using real money, here we consider a system where monetary transfers are prohibited. A user who wants to backup his data must in return supply some of his resources (storage space, upload and download bandwidth) to the system. We propose a hybrid P2P architecture where all backup data is transferred directly between peers, but a dedicated server coordinates all operations and maintains meta-data. We achieve high reliability guarantees while keeping our data replication factor low by adopting sophisticated erasure coding technology (cf., [2]).

The Market Design Problem. Using decentralized peers to store data also comes at a cost, raising two market design challenges regarding *incentives* and *efficiency*. Every user must provide a certain amount of all three resources, even if he currently only consumes one or two resources. Consequently, balancing a user's consumption and supply *per resource* does not make sense. Furthermore, it is natural that each user has different preferences regarding how much of each resource he wants to supply. Thus, a rigid accounting system that enforces the same resource ratios across all users is undesirable. Consequently, a sophisticated mechanism is necessary to first elicit users' preferences regarding their resources and then assign work to users in a way that maximizes overall efficiency.

Related Work. One of the early research projects investigating distributed file systems was Farsite [2]. However, in Farsite there were no incentives for users to contribute their resources. More recently, researchers have looked at the incentive problem and proposed market-based solutions (e.g., for computational

---

⋆ A long version is available at http://eecs.harvard.edu/~seuken/AMMA09_long.pdf

S. Das et al. (Eds.): Amma 2009, LNICST 14, pp. 55–57, 2009.

grid networks). The proposed solutions, however, generally require sophisticated users able to specify bids in an auction-like framework. The two papers most similar to our work are [1] and [3]. They analyze exchange economies for improving the efficiency of file-sharing networks. While the domain is similar to ours, however, the particular challenges they face are quite different.

2   Research Contributions

In this paper we make five key contributions regarding P2P backup systems:

1. **Design:** We present the complete design of a P2P resource exchange market including a server-based framework for decentralized resource supply and service consumption, an accounting system, and a work allocation method.
2. **User Interface:** We present a UI that hides the market from the user (see Figure 1). The design challenge was to provide market information to the user and to elicit the user's preferences with as little interaction as possible.
3. **Equilibrium Analysis:** We introduce the concept of a "buffer equilibrium", a desirable state of the system that is reached when the ratio of demand and supply is the same for all three resources. We formally prove existence and uniqueness of the buffer equilibrium under quite general assumptions.
4. **Price Update Algorithm:** We present a price update algorithm that uses supply and demand information to move prices towards the equilibrium. We prove analytically and show via simulations that the algorithm converges to a buffer equilibrium under some technical assumptions.
5. **Payment Mechanism:** We introduce a sophisticated payment mechanism that addresses various incentive problems that arise in practice.

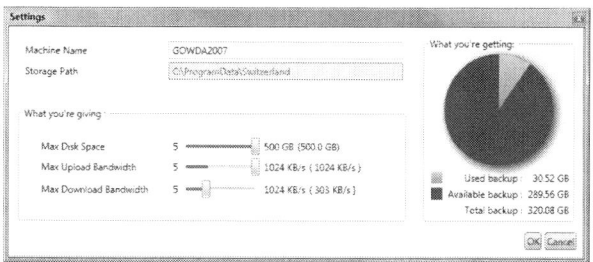

Fig. 1. Screenshot of the Current User Interface: The Settings Window

3   Future Work

The market design described in this paper is already implemented as part of a P2P backup system under development at Microsoft Live Labs. Thus, we are able to complement the theoretical analysis with discussions of implementation challenges. An alpha version of the software has already been released to Microsoft employees. We are currently collecting data on supply, demand and price developments over time to perform an empirical analysis of this market.

References

1. Aperjis, C., Johari, R.: A Peer-to-Peer System as an Exchange Economy. In: Proceedings from the workshop on Game theory for communications and networks (GameNets), Pisa, Italy (October 2006)
2. Bolosky, W.J., Douceur, J.R., Howell, J.: The Farsite Project: A Retrospective. SIGOPS Operating Systems Review 41(2), 17–26 (2007)
3. Freedman, M.J., Aperjis, C., Johari, R.: Prices are Right: Managing resources and incentives in peer-assisted content distribution. In: Proceedings of the 7th International Workshop on Peer-to-Peer Systems, Tampa Bay, Florida (February 2008)

School Choice: The Case for the Boston Mechanism

Antonio Miralles

Boston University, Dept. Economics, 270 Bay State Road, Boston MA02215 USA
Miralles@bu.edu

Since Abdulkadiroğlu and Sönmez's [3] work, a concern on the mechanisms used to assign children to publicly funded schools endures. Among other school districts, Boston has concentrated a lot of attention. The formerly called Boston Mechanism (BM) that was applied since 2000 has been widely criticized. Finally in 2005, the Boston Public School authority decided to replace this mechanism with the so-called Deferred Acceptance (DA) algorithm. The present paper argues that replacing BM might not be recommendable in every case, hence providing rationale to its persistence in other municipalities such as Cambridge, MA, Denver and Minneapolis.

Both in BM and in DA, parents are requested to submit a ranking with their ordinal preferences over schools. Both mechanisms follow then a multi-round assignment algorithm. In round 1, each student is considered for the school her parents put in first position in the ranking. The mechanism tries to assign this student to that school. There would be schools for which too many students are considered, and some would need to be rejected. In doing so, schools use some priority criteria (e.g. sibling already attending the school, geographical proximity, etc.) previously fixed by the school authority. Remaining ties are broken via fair lotteries. Rejected students go to the next round, and they are considered for the schools parents have ranked next in their lists. The process is repeated until a round is reached in which no student is rejected.

The difference between the two mechanisms concerns accepted students. In BM, an accepted student obtains a slot for sure at the school. Student and slot are not considered in further rounds. In DA, all students have to be reconsidered in further rounds as long as not everyone is accepted. Acceptances are meanwhile tentative. An accepted student would just be reconsidered for the same school in the next round.

This difference is paramount. DA is strategy-proof. It is weakly optimal for any parent to submit truthful reports no matter what other parents do. In BM, available slots diminish after each round. Thus it may be convenient to rank a less demanded school higher than what true ordinal preferences would induce. The fact that parents have to strategize leads to two main problems. One is lack of coordination (Ergin and Sönmez [5]). The other is that sophisticated parents may take advantage of non-strategic (naïve) parents (Pathak and Sönmez [7]), who may be harshly punished.

There is a new turn in this debate, which I subscribe. Erdil and Ergin [4] and Abdulkadiroğlu, Che and Yasuda [1] observe that DA leads to efficiency losses when priorities are weak (i.e. there are fairly less priority categories than students). This is usually the case in school choice. Lotteries are needed to break multiple ties. Therefore, parents face uncertainty when taking their decisions. Cardinal utilities, or preference intensities over schools, determine best responses. And DA, precisely due to its strategy-proofness, ignores any information on these intensities. In BM, parents with

S. Das et al. (Eds.): Amma 2009, LNICST 14, pp. 58–60, 2009.

identical ordinal preferences may submit different reports according to their intensities. BM distinguishes among them, allowing more information to be used.

To derive theoretical results, I analyze a simplified scenario with a continuum of perfectly informed rational (sophisticated) students, finitely many schools, and no priorities (i.e. all students belong to the same category). I obtain that DA performs very poorly if all the students share the same ordinal preferences over schools. Any other anonymous mechanism would weakly ex ante Pareto-dominate DA. I also obtain that BM achieves some ex ante efficiency properties that DA does not. I show that BM resembles (and sometimes coincides with) the Pseudomarket mechanism conceived by Hylland and Zeckhauser [6]. These results are robust to the inclusion of sibling priorities (for few students and highly correlated to ordinal preferences).

Nevertheless, BM punishes non-strategic parents. I suggest a partial solution for that problem. Abdulkadiroğlu, Pathak, Roth and Sönmez [2] observe that more than 35% parents in Boston in 2001 ranked an overdemanded school (one that has no slots available after the first round) in second position. Assuming that parents have reliable information on school demands, this way to proceed is naïve. If the student is rejected in the first round, she is definitely rejected in the second. This waste of one round may lead the student to a very bad assignment. I argue that this problem can be solved by erasing overdemanded schools from parents' lists at the end of each round, hence providing naïve parents with a partial protection device. This improvement is additionally innocuous if all parents are sophisticated.

I simulate a variety of scenarios: 4, 5 and 6 schools; 20, 30 and 40 slots per school; no priorities and walking-zone priority; correlation and no correlation of cardinal utilities with priority status; different levels of correlation of valuations across parents; no naïve students and half of them naïve; unprotected and partially protected naïve students. BM outperforms DA in utilitarian expected welfare in all cases. Welfare gains increase with the correlation of valuations among parents, and decreases with the valuation-priority correlation. Naïve students are harshly punished if they are not protected and there is high correlation among parents' valuations. However, the partial protective device is effective, and it does not always diminish utilitarian welfare gains.

DA shall not replace BM in all cases. BM has market-oriented efficiency properties which are important in real-case scenarios with coarse priorities. Instead, I propose BM to be partially modified so as to diminish the pervasive effects it may have over non-strategic parents. Improving BM the way I suggest has negligible implementation costs, as compared to the replacement policy once applied in Boston.

References

1. Abdulkadiroğlu, A., Che, Y., Yasuda, Y.: Expanding "Choice" in School Choice. Mimeo (2008)
2. Abdulkadiroğlu, A., Pathak, P.A., Roth, A.E., Sönmez, T.: Changing the Boston School Choice Mechanism: Strategy-proofness as Equal Access. Mimeo (2006)
3. Abdulkadiroğlu, A., Sönmez, T.: School Choice, a Mechanism Design Approach. American Economic Review 93, 729–747 (2003)
4. Erdil, A., Ergin, H.: What's the Matter with Tie-Breaking? Improving Efficiency in School Choice. American Economic Review 98, 669–689 (2008)

5. Ergin, H., Sönmez, T.: Games of School Choice under the Boston Mechanism. Journal of Public Economics 90, 215–237 (2006)
6. Hylland, A., Zeckhauser, R.: The Efficient Allocation of Individuals to Positions. Journal of Political Economy 87, 293–314 (1979)
7. Pathak, P., Sönmez, T.: Leveling the Playing Field: Sincere and Sophisticated Players in the Boston Mechanism. American Economic Review 98, 1636–1652 (2008)

Turing Trade:
A Hybrid of a Turing Test and a Prediction Market

Joseph Farfel and Vincent Conitzer

Duke University

Abstract. We present Turing Trade, a web-based game that is a hybrid of a Turing test and a prediction market. In this game, there is a mystery conversation partner, the "target," who is trying to appear human, but may in reality be either a human or a bot. There are multiple judges (or "bettors"), who interrogate the target in order to assess whether it is a human or a bot. Throughout the interrogation, each bettor bets on the nature of the target by buying or selling human (or bot) securities, which pay out if the target is a human (bot). The resulting market price represents the bettors' aggregate belief that the target is a human. This game offers multiple advantages over standard variants of the Turing test. Most significantly, our game gathers much more fine-grained data, since we obtain not only the judges' final assessment of the target's humanity, but rather the entire progression of their aggregate belief over time. This gives us the precise moments in conversations where the target's response caused a significant shift in the aggregate belief, indicating that the response was decidedly human or unhuman. An additional benefit is that (we believe) the game is more enjoyable to participants than a standard Turing test. This is important because otherwise, we will fail to collect significant amounts of data. In this paper, we describe in detail how Turing Trade works, exhibit some example logs, and analyze how well Turing Trade functions as a prediction market by studying the calibration and sharpness of its forecasts (from real user data).

Keywords: prediction markets, Turing tests, games with a purpose, deployed web-based applications, using points as an artificial currency.

1 Introduction

In a Turing test, a single human being (the *judge*) chats with two mysterious conversation partners [6]. One of the two mystery conversationalists is another human, while the other is a computer program (a *chat bot*, or just *bot*). The bot is the entity who is actually taking the test: If the judge cannot (accurately) tell which mystery conversation partner is the human and which is the bot, then the bot passes the test (and otherwise it fails). It is easy to see that a Turing test can also be run with only a single mysterious conversation partner (whom we will call the *target*). To do so, the test organizer chooses a human target with probability 50% and a computer target with probability 50%. Then, after a conversation with the target, the judge is asked to report how probable she thinks it is that the target is a human—if she reports 50% or higher when talking to a bot, then that bot passes the test.

S. Das et al. (Eds.): Amma 2009, LNICST 14, pp. 61–73, 2009.

One might imagine a variant of the Turing test where the "judge" consists of a *group* of humans (more generally, *agents*), and the result of the test hinges upon the group's aggregate belief of the probability that the target is a human. This setting involves multiple judging agents, where each agent has her own belief, but where information about beliefs might be exchanged among agents during the course of the test. An agent's personal belief is updated throughout the test based on the information she receives about other agents' beliefs, and, of course, on the target's contributions to the conversation.

Our new web game, Turing Trade, is an implementation of a Turing test with a group as the judge. In Turing Trade, a group of agents converses with a single target. Each individual agent in the group gets to ask the target public questions, and the target gives public answers. During the conversation, all individuals in the group are encouraged to competitively bet on the target's humanity, by buying and selling securities (with points, not real money). The price of these securities varies based on judges' bets, and at any given time in the game, this price is a measure of the group's consensus belief that the target is a human. At the end of the game, the target's true nature (human or computer) is revealed, and based on this some of the securities pay out. The betting part of the game is a *prediction market* [9], where the single binary event that the judges are trying to predict is "the target will be revealed to be a human."

Turing Trade can be played online at http://www.turingtrade.org. All logs from played games are posted publicly on the website. There are previously existing websites where one can take a more traditional Turing test, notably the Turing Hub, at http://www.turinghub.com, where a single player can log in as a judge, have a conversation with a target, and then rate the target's humanity on a four-point scale. One goal of Turing test web sites is to gather data from humans to help improve the conversation skills of bots. Having a large database of conversation logs, each with some attached humanity rating, would certainly be valuable for designing and training chat bots, and possibly for AI in general. We believe that Turing Trade has at least the following advantages over more traditional Turing test websites (such as the Turing Hub):

1. **Entertainment.** We believe that playing Turing Trade is more fun than participating in a normal Turing test. Apart from the social amusement provided by the interesting and clever questions submitted by other members of the judging group, the game encourages competition, by rewarding judges who increase the accuracy of the consensus probability estimate.

2. **More data, from more volunteer judges.** *Games with a purpose* use entertainment value to convince legions of humans to do something useful that is (currently) difficult for computer programs [8]. For example, playing the ESP Game, at http://www.espgame.org, is a fun way to help put useful labels on all of the images on the web [7]. In the first four months of the ESP Game's existence, 13,630 people played the game (over 80% of which played on more than one occasion), and an informal recent check of the website at various times of day implies that about 40 people are playing the game at any given moment. Very few people (certainly, fewer than the numbers mentioned) would sit and label images without compensation if it were not in the context of a game (with competition, cooperation, points,

etc.). Similarly, Turing Trade's goal is to attract more Turing test judges (and human targets) than its non-game contemporaries.

3. **Better data, through proper incentives.** Most Turing test web sites offer no incentives to the judge. Even if the mere act of having a conversation with a mysterious subject is incentive enough for people to participate in the test, a judge is certainly not strictly incentivized to report her truthful belief about the nature of the target at the end. Turing Trade's prediction market betting system incentivizes a bettor to bet in a way reflecting the true probability she assigns to the target being a human; moreover, it encourages the bettor to improve her own acuity at estimating this probability, by punishing those who predict incorrectly, and rewarding those who predict correctly. (Punishments and rewards are in the form of points, rather than real money, but this is better than no incentive at all, and in fact the lack of real money does not seem to greatly affect the accuracy of a prediction market [5].)

4. **More mystery.** Judges having conversations at the Turing Hub are immediately biased toward thinking that they are speaking to a bot: since the site has only light traffic, the chances of a human-human conversation are quite low, and to make things worse, some of the bots on the site use custom (and very bot-like) message windows. More player traffic (combined with the ability to play as a target), as well as a consistent interface, causes Turing Trade's targets to be more mysterious (which is also more enjoyable).

5. **Fine-grained data.** In Turing Trade, a group's current consensus evaluation of the probability that the target is a human is given by the current price of the securities. Since this price varies over the course of a conversation, our data not only gives an overall assessment of how human-like a target acted in a particular conversation, but also shows how the impression that the target made varied over time. For example, a game log might show that the security price stayed high for a while, and then dropped sharply after the target answered question 5. This would imply that the target gave human-like answers to questions 1-4, but not to question 5. One can imagine mining mountains of logs for sharp price drops and rises, thereby compiling lists of good questions, as well as good and bad answers to them. This should help in the design of better chat bots as well as in the training of judges. We provide some examples of log data generated by Turing Trade in Section 4.

Apart from web-based Turing tests like those at the Turing Hub, there are a few regular Turing test-based competitions, some offering cash prizes to the most human-like participating bot. The most famous of these is the Loebner Prize, which claims to be the first formal instantiation of a Turing test (http://www.loebner.net/Prizef/loebner-prize.html). This yearly competition, started in 1990, features four judges (usually university professors), each of whom scores every entering bot. The Loebner Prize offers a $100,000 grand prize and a solid gold medal to the first bot whose responses are "indistinguishable from a human's." Although this prize goes unclaimed, an annual prize of $2,000 is offered to the most human-like bot in the competition.

Though they work fine as Turing tests, and are good indicators of which bots are currently the most advanced, the Loebner Prize competition and other competitions like it do not serve the same purpose as Turing Trade. Turing Trade's purposes include: (1) to collect large quantities of fine-grained data for use by bot designers, (2) to introduce a novel, fast-paced prediction market, which may provide valuable lessons for the design of other prediction markets, and (3) to provide entertainment value.

2 Game Overview

In a game of Turing Trade, the *target* is the single player (possibly a bot) whose humanity is being judged. The group judging the target is composed of n agents, called *bettors*. The bettors ask the target questions, and bet on whether the target is a human or a computer. The target answers questions from the bettors, and tries to seem as human as possible, whether or not it is really a human.

As a brief note on implementation, all players in the game (bettors and target) communicate through web-based Java applets. These applets send all information through a central server, also written in Java. The server is capable of managing multiple simultaneous Turing Trade games. In the current incarnation of the game, the number of bettors, n, is restricted to three, at most, per game (this is not due to scalability reasons but rather to ensure that all bettors have a chance to ask questions).

2.1 Bot Targets

It is very important for our game to have a strong lineup of bots available to serve as targets. The bots described below (except for Simple Bot) were written by others, and reside on their owners' web servers (it is not our intention to create new bots ourselves). When a game is in progress, the Turing Trade server initiates a new conversation with a bot, and simply sends it bettors' questions and receives the bot's answers. The current incarnation of the game features six different bots organized into three classes:

1. **Simple Bot.** This is a very simple bot—to any question, it replies with the same answer ("Hmmm... That's an interesting question.").
2. **Alice and iGod.** These bots are based on AIML, or the Artificial Intelligence Markup Language. AIML and Alice, the first bot to use it, are creations of Dr. Richard Wallace (http://www.alicebot.org); they are extensions of the logic underlying the classic bot Eliza, developed by Joseph Weizenbaum in 1966. The iGod bot is currently the most popular bot at the free AIML-bot hosting web site Pandorabots (http://www.pandorabots.com/). Alice won the Loebner Prize for most human-like chat bot in 2000, 2001, and 2004.
3. **Jabberwacky, George, and Joan.** These three bots are all based on Jabberwacky, by Rollo Carpenter (http://www.jabberwacky.com/). Its approach is heavily centered on learning, and it operates primarily by storing everything that any human has ever said to it, and using contextual pattern matching to find things in this vast database to say in response to new human input. Given this emphasis on learning, Jabberwacky-based bots are especially good candidates for using Turing Trade logs (which include not only a conversation log, but also information about the varying

perception of the target's humanity during the conversation) to improve performance. The Jabberwacky bots George and Joan won the Loebner prize in 2005 and 2006.

2.2 Questions and Answers

The target always sees only one question at a time from the group of bettors. This question is called the *current* question. The target considers the current question, and sends its answer to the server; at this point, the server may send the target another single (current) question. The target may only send one answer for each question. From the target's perspective, the conversation is a simple back-and-forth exchange. The only indication the target has that it is talking to a group of people rather than a single person is that questions belonging to bettor i are tagged as such. This allows chat bots that currently exist to play Turing Trade unmodified.

Every bettor also gets to see the current question, at the same time as the target. Unlike the target, however, which may only submit an answer if there is an unanswered current question, any of the n bettors may submit a question to the game server at any point during the game. The server keeps a queue of questions, Q_i, for each player i, and initially, all question queues are empty. When the server receives a question, q, from a bettor i, it does the following:

- If there is no current (unanswered) question, broadcast question q to all bettors and the target. Question q is now the current question.
- Otherwise (there is a current question), add the question q to queue Q_i.

With this setup, all bettors and the target see the current question, but every other unanswered question is invisible to everyone but the bettor who asked the question. When the server receives an answer from the target to a current question q, it does the following:

1. Let i be the bettor who asked the question being answered (question q). Send the answer to bettor i, and send a signal (not containing the text of the answer) to every bettor $b \neq i$ signifying only that an answer to the question has been given.
2. Starting with $j = i + 1 \pmod{n}$, and incrementing $j \pmod{n}$ after each check, search for the first nonempty queue Q_x.
   - If all queues are empty, do nothing (except wait for a bettor to send a question).
   - Otherwise (Q_x is the first nonempty queue), remove the first question q' from Q_x. This is the new current question; send q' to all bettors and the target.
3. Five seconds after step 1 (sending the answer to question q to player i), send the text of the answer (to q) to every bettor $b \neq i$.

This scheme ensures that questions are taken from bettors in a round-robin manner, unless some bettors are not asking questions, in which case they are skipped. We note that bettor i gets to see the answer to her question five seconds earlier than every bettor $b \neq i$. This delay rewards bettors for asking good questions (where a good question is one that reveals a lot about the nature of the target), because it allows the bettor who asked the question to trade on this information before it becomes available to the other bettors. An example of a bettor's view of an in-game conversation is shown in Figure 1.

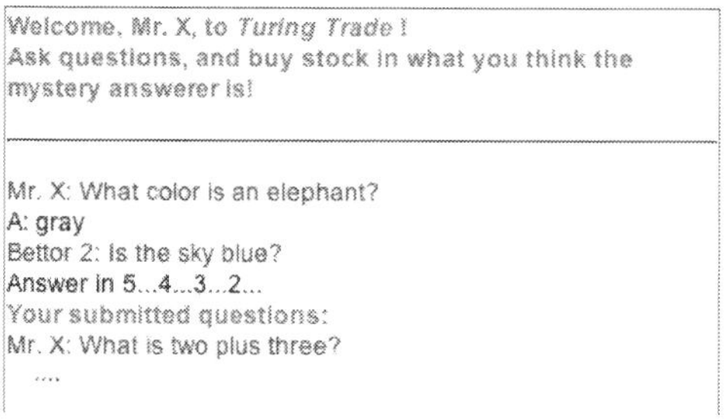

```
Welcome, Mr. X, to Turing Trade !
Ask questions, and buy stock in what you think the
mystery answerer is!
_____

Mr. X: What color is an elephant?
A: gray
Bettor 2: Is the sky blue?
Answer in 5...4...3...2...
Your submitted questions:
Mr. X: What is two plus three?
    ....
```

Fig. 1. An example of a bettor's ("Mr. X") in-game view of a conversation between the bettors and the target. This bettor owns the current question, shown at the bottom. She is also about to see the answer to Bettor 2's question, which Bettor 2 saw four seconds ago.

2.3 The End of the Game

A game of Turing Trade ends in one of several ways:

- **Time runs out.** Each game is timed, and the time limit is fairly short (in the current incarnation, it is two minutes). While Alan Turing hypothesized that machines in the year 2000 with 119 MB of memory would be able to regularly pass a five-minute Turing test [6], this prediction certainly did not come to pass. Our empirical results show that bettors usually (but not always) become extremely certain of a target's nature even before the end of two minutes.
- **All bettors signify that they are done betting.** If a bot (or human) gives a particularly elucidating answer or two, bettors may become virtually certain of the target's nature. We give bettors the option to signify that they are satisfied with their current bets, and wish to end the game early. All bettors must agree to end early.
- **The target (or all of the bettors) leaves the game early.**

At the end of a game, the nature of the target is revealed to all of the bettors. The bettors and the target are rewarded based on the bettors' bets and the target's true nature.

2.4 Betting

At any time during the conversation with the target, any bettor may place a bet on whether or not the target is human. A bet is made by buying or selling a *human security* or a *computer security*. A human security is an asset of the form "Pays 100 points if the target is revealed to be a human," while a computer security is an asset of the form "Pays 100 points if the target is revealed to be a computer." Securities pay out at the end of the game, when the target's nature is revealed to bettors: for example, if the target is a human, a human security pays out 100 and a computer security pays out 0. Since the two types of security are complementary (owning one of each type of security is equivalent to owning 100 points), we without loss of generality restrict every bettor to own at most one type of security at a time.

Fig. 2. The interface a bettor uses to buy and sell human and computer securities. The pictures indicate "betting human" (which means either buying human securities, or selling computer securities), and "betting computer" (which means either selling human securities, or buying computer securities). The number of securities the bettor owns, and the securities' type, is shown below the buttons. If the security price reaches a steady (and boring) equilibrium (usually at 100 or 0), a bettor can click "done buying;" if all bettors do this, the game ends early.

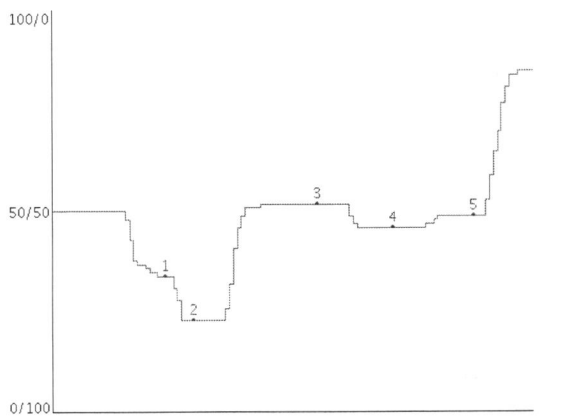

Fig. 3. An example of the price graph shown to the bettors and the target during a game of Turing Trade. Time progresses left to right, while the y-axis shows the price to purchase a human security. Dots indicate when the target answered a question. As one might expect, large shifts in the security price usually occur shortly after bettors see an answer (and update their impression of the target accordingly), while the price reaches equilibrium between answers.

Human and computer securities are bought from and sold to a central *market maker*, who has an infinite supply of securities to sell, and an infinite willingness to buy securities. The market maker always sets the price for the computer security at 100 minus the price of the human security (this is ignoring a small bid-ask spread that we will discuss shortly). A bettor can purchase or sell one security at a time. When a human security is purchased (or a computer security is sold), the price for human securities increases by 1, and when a human security is sold (or a computer security is purchased) the price for human securities decreases by 1.

The market maker maintains a spread of 1 between bid and ask prices. This is done to prevent arbitrage: with the spread, a bettor can buy a security for the ask price of x from the market maker (causing the ask price to increase to $x + 1$), and then sell it back for the bid price $(x + 1) - 1 = x$. Neither the bettor nor the market maker profits if this happens, but without the spread, the bettor would have had a profit of 1. The maximum price for a human security is 100, and the minimum price is 0. An example of the interface that a bettor uses to buy and sell securities is shown in Figure 2.

The price to buy a human security is plotted over the course of a game (an example is shown in Figure 3). Local equilibria in the price measure the bettors' consensus belief (at some point in time) of how probable it is that the target will be revealed to be a human. For example, if the human security price is hovering around 70, then the bettors, in aggregate, believe that the there is a 70 percent chance that the target will be revealed to be a human at the end of the game. This interpretation relies upon the assumption that the bettors are rational (in the sense of maximizing their expected number of points), and upon the fact that a prediction market such as this one offers incentives for rational bettors to update the consensus probability in ways consistent with their true beliefs (at least for a myopic sense of rationality).

3   Evidence for the Accuracy of Prediction Markets

Empirical evidence has shown that prediction markets are quite good at forming accurate probability estimates for events. For example, the Iowa Electronic Markets outperformed 451 out of 591 major public opinion polls in predicting the margin of victory in past U.S. presidential elections [1]. Perhaps surprisingly, even markets using play money exhibit very strong predictive powers. Pennock *et al.* discovered high prediction accuracy both for the Foresight Exchange (http://www.ideosphere.com), where traders bet on the outcomes of open scientific questions, and for the Hollywood Stock Exchange (http://www.hsx.com), where ending security prices for Oscar, Emmy, and Grammy awards were found to correlate well with actual award frequencies [3]. Similarly, Servan-Schreiber *et al.* found no statistically significant difference in the accuracy of play money and real money prediction markets in predicting the outcomes of American Football games during the 2003-2004 NFL season [5]. Results like these bode well for the accuracy of Turing Trade's prediction market. This is especially promising because as bots improve their conversation skills, and become less distinguishable from humans, judges' predictions will need to become more accurate to detect what subtle differences remain. In Section 5, we assess the predictive powers of Turing Trade directly, based on real data.

4   Example Logs

Figures 4, 5, and 6 contain excerpts of some real logs from Turing Trade. The graph at the top of each log shows time on the x-axis, and the human security price on the y-axis. The log detailing the game times at which events happened appears below. Logs have been edited so that they show all questions and still fit within the space limits (this involved removing large chunks of entries detailing each new computer or human bet; some entries describing players joining or leaving the game have also been removed).

Fig. 4. This run demonstrates the ability of bots to sometimes evade conclusive detection for an extended period during a game. In this run, the Jabberwacky-based bot George was able to seem at least somewhat human, until its ridiculous answer to question 5.

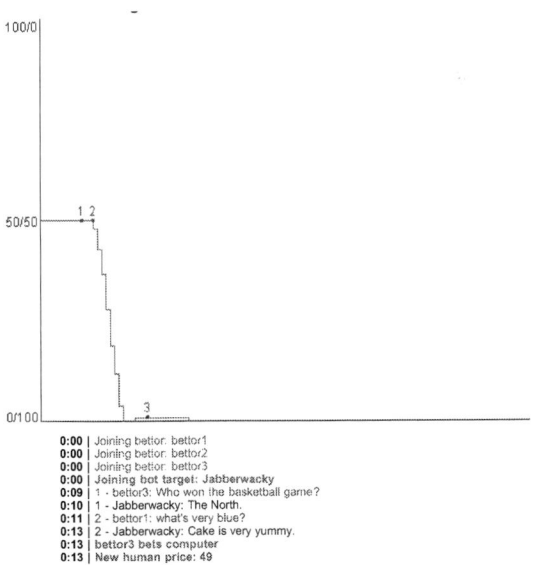

Fig. 5. Sometimes, a bot will seal its fate with its very first answer

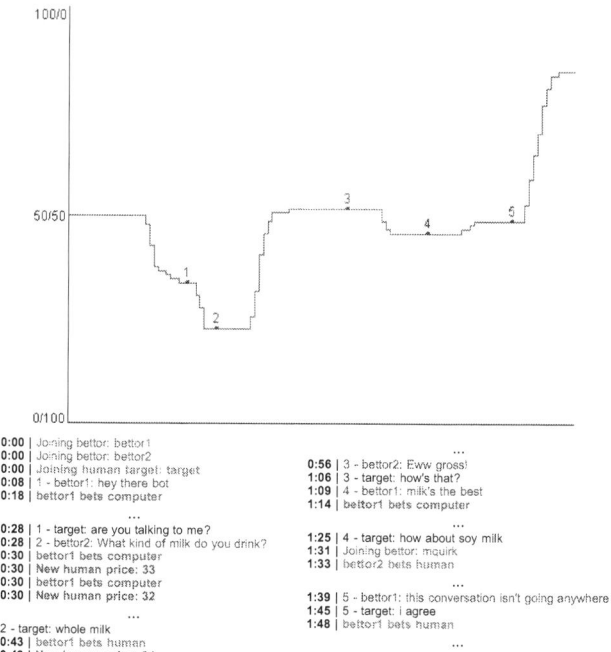

Fig. 6. A human target. This log is an example of how a human can sometimes seem like a computer, even when his answers to questions are perfectly reasonable.

5   Calibration and Sharpness

How can we evaluate whether our prediction market is functioning well? One desirable property is that the predictions are *calibrated*. This means the following. Suppose we consider all the runs where, after a given amount of time, the market probability (price) that the target is human is at (say) 10%. We would hope that in exactly 10% of these runs, the target is indeed a human. If this is true for all probabilities, then the market is (perfectly) calibrated.

A practical problem with this definition is that we generally do not have many data for each individual probability. To address this, it is common to bin the probabilities together. For example, we consider all the runs where, after a given amount of time, the market probability of a human target is between 10% and 20%, and ideally the fraction of these runs where the target is indeed a human is between 10% and 20%. In practice, even this is often not the case for every bin, but we would hope that the market probabilities and the true fractions are at least close.

Currently, about 900 games of Turing Trade have been played. After removing the logs from games where no bets were made, we examined the remaining 694 game logs to determine market calibration. Figure 7 illustrates the results. The market seems reasonably, albeit not perfectly, calibrated. One would expect that the market would

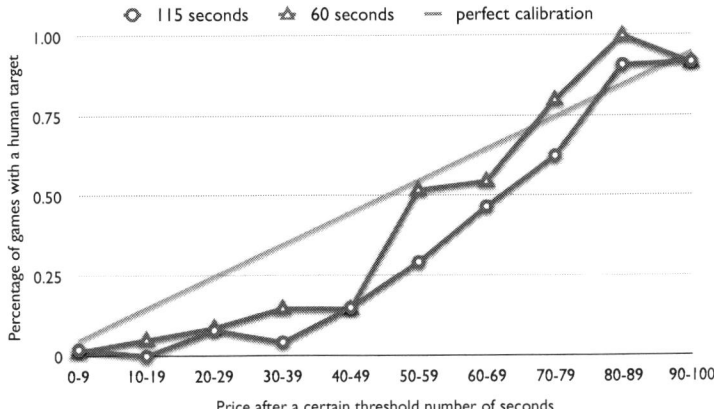

Fig. 7. The x-axis denotes bins of runs, partitioned by the human security price after a given number of seconds. The y-axis is the fraction of runs in a bin that had a human target. If the market were perfectly calibrated, these would match.

become even more calibrated over time, especially as players accrue more experience and become better bettors.

To have a good prediction market, it is not sufficient that it is calibrated. For example, suppose it is known that 50% of targets are human, and the initial market probability is always 50%. Then, if traders never trade at all, the prediction market is perfectly calibrated—but this would constitute a completely dysfunctional prediction market. The missing property is that of *sharpness*: we want the market predictions to be close to

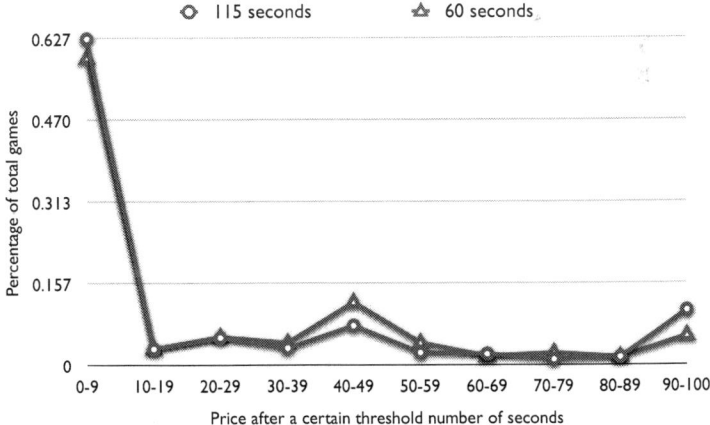

Fig. 8. The x-axis denotes bins of runs, partitioned by the human security price after a given number of seconds. The y-axis is the fraction of total runs in a bin. The high percentages at the lowest and highest bins indicate sharpness, while the spike at the $40 - 49$ bin is undesirable.

0% or 100%. As it turns out, the Turing Trade prediction market makes very sharp predictions, as illustrated by Figure 8.

6 Conclusions

We introduced a new "game with a purpose" called Turing Trade. The game is a group-judged Turing test, where members of the judging group (bettors) bet on whether their mystery conversation partner (the target) is a human or a computer. Betting is accomplished through the use of a prediction market, with bettors using play money to buy and sell "human securities" and "computer securities" from an automated market maker. We believe the game offers numerous advantages over standard Turing test websites, including the promise of collecting significantly more data, and more finely-grained data, for chat bot designers and others. The Turing Trade project has additional purposes, including the creation of a novel, fast-paced prediction market that may provide useful lessons for the design of prediction markets in general. Another purpose is simply to create entertainment value for its players. If the game ends up being played by very many people, then, for example, providers of free e-mail accounts could use our game to ensure that bots do not sign up for accounts (a trick commonly used by spammers), by requiring a new user to play as the target in Turing Trade (and be judged human).

Prediction markets have worked very well, empirically, in other settings [1], even when fake money is used (as is the case with Turing Trade) [3,5]. We have already obtained a significant number of runs with our publicly available web-based implementation of Turing Trade (http://turingtrade.org), especially after it received some attention on blogs including http://www.midasoracle.org and http://www.marginalrevolution.com. Our analysis of these runs suggests that Turing Trade produces very strong and quite accurate predictions after short periods of time, with the market price responding rapidly to good or bad answers by the target.

Acknowledgments

We thank the National Science Foundation for support under award number IIS-0812113, the Alfred P. Sloan Foundation for support under a Research Fellowship, and Yahoo! for support under a Faculty Research Grant. We also thank all the people that have played Turing Trade and have given us valuable feedback.

References

1. Berg, J., Forsythe, R., Nelson, F., Rietz, T.: Results from a Dozen Years of Election Futures Markets Research. In: Handbook of Experimental Economics Results (2001)
2. Hanson, R.: Logarithmic market scoring rules for modular combinatorial information aggregation. Journal of Prediction Markets 1, 3–15 (2007)
3. Pennock, D.M., Lawrence, S., Giles, C.L., Nielsen, F.A.: The real power of artificial markets. Science 291, 987–988 (2002)
4. Savage, L.J.: Elicitation of personal probabilities and expectations. Journal of the American Statistical Association 66, 783–801 (1971)

5. Servan-Schreiber, E., Wolfers, J., Pennock, D.M., Galebach, B.: Prediction markets: Does money matter? Electronic Markets 14 (September 2004)
6. Turing, A.: Computing machinery and intelligence. Mind 59, 433 (1950)
7. von Ahn, L., Dabbish, L.: Labeling images with a computer game. In: Proceedings of the SIGCHI conference on Human factors in computing systems, pp. 319–326 (2004)
8. von Ahn, L., Dabbish, L.: Designing Games with a Purpose. Communications of the ACM 51, 58–67 (2008)
9. Wolfers, J., Zitzewitz, E.: Prediction Markets. The Journal of Economic Perspectives 18(2), 107–126 (2004)

A Market-Based Approach to Multi-factory Scheduling

Perukrishnen Vytelingum[1], Alex Rogers[1], Douglas K. Macbeth[2], Partha Dutta[3],
Armin Stranjak[3], and Nicholas R. Jennings[1]

[1] School of Electronics and Computer Science, University of Southampton, UK
{pv,acr,nrj}@ecs.soton.ac.uk
[2] School of Management, University of Southampton, UK
D.K.Macbeth@soton.ac.uk
[3] Strategic Research Centre, Rolls Royce, UK
{Partha.Dutta,Armin.Stranjak}@Rolls-Royce.com

Abstract. In this paper, we report on the design of a novel market-based approach for decentralised scheduling across multiple factories. Specifically, because of the limitations of scheduling in a centralised manner – which requires a center to have complete and perfect information for optimality and the truthful revelation of potentially commercially private preferences to that center – we advocate an informationally decentralised approach that is both agile and dynamic. In particular, this work adopts a market-based approach for decentralised scheduling by considering the different stakeholders representing different factories as self-interested, profit-motivated economic agents that trade resources for the scheduling of jobs. The overall schedule of these jobs is then an emergent behaviour of the strategic interaction of these trading agents bidding for resources in a market based on limited information and their own preferences. Using a simple (zero-intelligence) bidding strategy, we empirically demonstrate that our market-based approach achieves a lower bound efficiency of 84%. This represents a trade-off between a reasonable level of efficiency (compared to a centralised approach) and the desirable benefits of a decentralised solution.

1 Introduction

Job-shop scheduling is an important and challenging problem that has long been the subject of extensive research in Artificial Intelligence [2]. Specifically, it is the problem of allocating resources for the completion of customers' jobs within specific deadlines in a single or a number of factories. Such a problem is usually solved optimally as a combinatorial optimisation problem (that maximises the utility of all stakeholders, i.e. both customers and factories) by a center that has complete information of the system. However, this centralised approach can be problematic in a number of ways. First, a completely new solution often needs to be recomputed from scratch with every change in the system (e.g. new stakeholders entering the system or existing ones updating their job requirements). Second, stakeholders, which usually represent competing organisations in the real world, are often reluctant to share their private and sensitive information with a center that would then have complete knowledge of the whole system, since this may shade their competitive edge. Finally, and perhaps most importantly, the solution

S. Das et al. (Eds.): Amma 2009, LNICST 14, pp. 74–86, 2009.

quickly becomes intractable with increasing problem size, because of the combinatorial nature of the task at hand.

For these reasons, there is an increasing trend towards solving these scheduling problems in a decentralised manner. Such approaches are inherently more robust because they don't have a single point of failure and they also do not require the divulging of commercially valuable information to a third party. Moreover, the distributed nature of the computation also means a more scaleable solution is possible, with the system being more dynamic in adapting to changes in the system.

Against this background, this paper reports on the development of a market-based approach for scheduling across multiple factories. Our work is principally motivated by the need for more robust and scaleable solutions in the Aero Repair and Overhaul (AR&O) context where customers (typically airlines) in the system require that their engines be scheduled for routine maintenance in *overhaul bases* (OHBs). In addition, engines may require more urgent inspection due to some damage (e.g. through bird-strike or icing) or mechanical failure. These disruptions introduce uncertainties in the system which increase the complexity of the scheduling process since it requires us to dynamically construct schedules capable of effectively coping with such unforeseen events in real time. The challenge is to dynamically schedule repairs and routine maintenance subject to these constraints. To date, the state of the art consists of a pragmatic scheduling solution [8] that consists of a center that computes the engine repair schedules based on the capacity and capability of the multiple factories and the engine repair severity. However, this is a centralised solution that suffers from all the aforementioned problems.

In more detail, the AR&O scheduling problem, as described above, is an important class of problems in its own right (being worth over $25B worldwide in 2008). It also has characteristics that are found in many other applications as diverse as classic job-shop scheduling, production planning and manufacturing scheduling [2,6]. Thus, in addition to its immediate goals of addressing the AR&O problem, this research endeavour has further applications within many adaptive decision processes where there is an extended and dynamic network of interdependent customer and supplier business entities. As such, it has the potential to provide utility to many other supply network settings where again traditional approaches have tended to a static representation unable to respond quickly to the rate of change in the environment.

To address the challenge of developing an agile and decentralised scheduling system, a number of researchers have advocated the use of economic metaphors by adopting a market-based approach where customers and factories are self-interested agents[1] that compete, through offers to buy and sell resources, in order to maximise their utility. Markets are a particularly suitable approach in this context, because of their ability to facilitate resource allocation often with public information exchange and their distributed nature with the resource allocation emergent from the competition among buyers and sellers. Because of their distributed nature, no single agent computes the resource allocation, markets are robust against failure and, furthermore, markets have been shown to dynamically and efficiently react to changes in the system [3]. Now, there has been

---

[1] Self-interested agents are reluctant to share private information and are driven by the objective of maximising their own profit.

previous work that adopted a market-based approach in this domain. Specifically, Baker's work looked at a market-driven contract net mechanism to schedule a factory [1], and Rassenti *et al.* developed a sealed-bid combinatorial auction mechanism for scheduling flight landing and take-off [7]. However, while these mechanisms reduce the complexity of the particular problems they are tackling, they still need a center that collects all offers (which are not made public) to match them for scheduling. In contrast, Wellman *et al.* examine a number of auction mechanisms for decentralised scheduling [12], including multiple simultaneous ascending auctions. Here, while the latter approach circumvents the need for an auctioneer (as all offers are made public in the market), it considers only scheduling for a single factory, and this solution does not easily generalise to the multi-factory case considered here because they consider fundamentally different auction mechanisms, namely single-sided ones where only buyers compete for resources.

As a consequence of the fundamental issues discussed above, the aforementioned market mechanisms fail to solve our motivating problem for decentralised AR&O scheduling. Thus, we address these issues by proposing a variant of one of today's most prominent auction formats, namely the Continuous Double Auction (CDA) [3], which allows multiple customers and multiple factories to compete in a market. In so doing, we extend the single-factory job-shop scheduling problem proposed by Wellman *et al.* to a more general one that allows multiple factories (OHBs in the AR&O context) to compete. Second, we design a variant of the traditional CDA that also considers the time factor of the scheduling problem. Thus, in more detail, this work extends the state of the art in the following ways:

- First, we develop a variant of the Continuous Double Auction for multi-factory scheduling. Our market-based approach is novel in being the first auction mechanism that allows multiple customers and multiple factories to compete in a market for scheduling jobs without the need for a center and without the need to reveal private and often sensitive information to a center.
- Second, we demonstrate the effectiveness of our approach by evaluating our market mechanism. Specifically, we provide a lower bound efficiency of 84%, with our market-based approach sacrificing at most 16% for the added benefits of a more robust, dynamic and transparent solution.

The remainder of this paper is structured as follows. We begin in Section 2 by formalising the multi-factory scheduling problem. In Section 3, we describe our mechanism which we empirically evaluate in Section 4. Section 5 concludes.

2 The Multi-factory Scheduling Problem

In this section, we describe the general scheduling problem that this work focuses on. Although motivated by our specific AR&O problem, we believe this applies to a broad class of domains. We first extend Wellman *et al.*'s scheduling model to deal with the problem of multi-factory scheduling, rather than the restrictive single-factory case that they consider. To this end, we consider several factories, potentially owned by different organisations, with the same number of one-hour unscheduled time-slots[2], denoted as $T = \{T_{start}, ..., T_{end}\}$. These time-slots can be allocated for customers' jobs, with each one having a limit price[3] representing the minimum price the factory will accept in exchange for that time-slot.

Next, we assume that each customer i, requires a single job of a certain *length* q^i (i.e. a number of time-slots), *value* ℓ^i (i.e the utility for all the q^i time-slots required) and *deadline* $t^i_{deadline}$ (i.e. completion is no later than a time-slot at $t^i_{deadline}$) scheduled in a single factory to follow up from Wellman *et al.*'s model. The customer is willing to spend no more than its value to have its job scheduled within its deadline. Furthermore, we assume that the customer has an inelastic demand[4]. That is, its utility is 0 if it cannot acquire sufficient time-slots to complete its job within its deadline.

The problem at hand is then to allocate a set of jobs in the available time-slots of a set of factories, subject to the length, value and deadline constraints of the jobs and the limit price constraint of the time-slots. In this context, given a set S_{cus} of customers and a set S_{fac} of factories, we have the demand, $demand^i$ $\forall i \in S_{cus}$ and the supply, $supply^{j,a}$ $\forall j \in S_{fac}$, $\forall a \in A^j$, where A^j groups time-slots with same limit prices given by:

$$demand^i = (id^i_{cus}, \ell^i, q^i, t^i_{deadline}) \; \forall i \in S_{cus}, \; t^i_{deadline} \in T \tag{1}$$

$$supply^{j,a} = (id^j_{fac}, c^{j,a}, f^{j,a}_{T_{start}}, ..., f^{j,a}_{T_{end}}) \tag{2}$$

where

$$f^{j,a}_t \in \{0,1\} \; \forall a \in A^j, \; \forall j \in S_{fac}, \; c^j_t = \sum_{a \in A^j} c^{j,a} f^{j,a}_t$$

and id_{cus} and id_{fac} are unique identifiers for customers and factories respectively, $\sum_{a \in A^j} f^{j,a}_t$ defines if time-slot t in factory j is allocated or not, $c^{j,a}$ is the limit price for a group of time-slots and c^j_t is the limit price of a time-slot t in factory j. Note that A^j can be defined over the space between a single set with all time-slots having the same limit prices and $|T|$ sets of single-time-slots in the case of different limit prices for all the time-slots of the factory[5].

We can now formalise the scheduling problem as a maximisation of profits of all stakeholders (to conform to the literature on the classic job-shop scheduling). We first define the following terms:

---

[2] Our choice of one-hour time-slots is not crucial to this work and, indeed, can be changed to represent time at arbitrary levels of granularities. In addition, there is no requirement that each factory has the same number of time-slots. We simply make these choices to simplify the simulations that follow later.

[3] The limit price corresponds to the production cost within that time-slot. A factory would allow usage of its time-slot only if a customer pays more than the associated production cost such that it does not make a loss.

[4] Note that the assumption of inelastic demand does not constrain our work. A customer with elastic demand would simply split its job into a set of q^i single-time-slot jobs.

[5] Here, $\sum_{a \in A^j} \sum_{t \in T} f^{j,a}_t = |T|$ and $\sum_{a \in A^j} f^{j,a}_t \leq 1 \; \forall t \in T, \; \forall j \in S_{fac}$. See Figure 1 for an example, with $|A^1| = 2$, $|A^2| = 4$ and $|A^3| = 2$).

– $\mathcal{A}(i,j) \in \{0,1\} \; \forall j \in S_{fac}, i \in S_{cus}$ specifies which customer is allocated to which factory.

– $\mathcal{TS}(i,j,t) \in \{0,1\} \; \forall j \in S_{fac}, i \in S_{cus}$ specifies which factory's time-slot is allocated to which customer.

The system is optimally scheduled when the following objective (profits of all stakeholders) is maximised, as we assume that stakeholders are self-interested, profit-motivated economic agents in the system. To this end, we must find:

$$\max \sum_{j \in S_{fac}} \sum_{i \in S_{cus}} \left[\mathcal{A}(i,j)\ell^i - \sum_{t \in T}(\mathcal{TS}(i,j,t)c_t^j) \right] \tag{3}$$

subject to the following constraints:

1. *Job deadline constraint (i.e. jobs are allocated within their deadline):*

$$\sum_{j \in S_{fac}} \sum_{t=T_{start}}^{t_{deadline}^i} \mathcal{TS}(i,j,t) = \sum_{j \in S_{fac}} \mathcal{A}(i,j)q^i, \; \forall i \in S_{cus}$$

2. *Factory's limit price constraint (i.e. all accepted jobs have value equal to at least the limit price for all time-slots):*

$$\sum_{t=T_{start}}^{t_{deadline}^i} \mathcal{TS}(i,j,t)c_t^j \leq \mathcal{A}(i,j)\ell^i, \; \forall i \in S_{cus}, \forall j \in S_{fac}$$

3. *Factory's time-slot scheduled to a single customer (i.e. no time-slots can be shared for more than one job):*

$$\sum_{i \in S_{cus}} \mathcal{TS}(i,j,t) \leq 1, \; \forall t \in T, \forall j \in S_{fac}$$

4. *Job scheduled to a single factory (i.e. factories cannot share jobs[6]):*

$$\sum_{j \in S_{fac}} \mathcal{A}(i,j) \leq 1, \; \forall i \in S_{cus}$$

5. *Customer's inelastic demand (i.e. jobs cannot be partially allocated):*

$$\sum_{j \in S_{fac}} \sum_{t \in T} \mathcal{TS}(i,j,t) = \sum_{j \in S_{fac}} \mathcal{A}(i,j)q^i, \; \forall i \in S_{cus}$$

Given complete and perfect information (with all agents truthfully revealing their preferences to a center), the center can optimally compute the solution to this problem (e.g. using the ILOG CPLEX optimisation tool). To this end, Figure 1 gives the demand and supply in the system along with the optimal solution (i.e. the allocated customers in each factory) to an example of such a scheduling problem with 3 factories and 8 customers. In particular, Customers 2, 6 and 8 are scheduled in Factory 1, Customers 4 and 5 in Factory 2 and Customer 1 in Factory 3.

---

[6] Note that this is certainly the case within the AR&O domain since engines are always serviced solely at one overhaul base. However, this might not generally be true, and depending on the specific setting, it is possible to relax this constraint.

Customer 1		Customer 5
•Value = $22		•Value = $3.5
•Length = 4		•Length = 1
•Deadline = 12		•Deadline = 9

Time	Factory 1		Factory 2		Factory 3	
	(with optimally scheduled customers in bold and unallocated time-slots denoted by x)					
9:00	$3.5	2	$2.25	5	$1.5	1
10:00	$3.5	2	$2.25	4	$1.5	1
11:00	$3.5	8	$4.5	x	$1.5	1
12:00	$3.5	6	$4.5	x	$1.5	1
13:00	$3.0	6	$4.5	x	$4.0	x
14:00	$3.0	8	$4.5	x	$4.0	x
15:00	$3.0	2	$3.25	4	$4.0	x
16:00	$3.0	8	$2.25	4	$4.0	x

Customer 2
•Value = $14.5
•Length = 3
•Deadline = 15

Customer 6
•Value = $9.5
•Length = 2
•Deadline = 16

Customer 3
•Value = $10.5
•Length = 3
•Deadline = 11

Customer 7
•Value = $3.25
•Length = 1
•Deadline = 9

Customer 4
•Value = $15
•Length = 3
•Deadline = 16

Customer 8
•Value = $12.75
•Length = 3
•Deadline = 16

Fig. 1. Multi-factory Scheduling Problem with 8 customers (with their limit price ℓ^i, length q^i and deadline $t^i_{deadline}$ $\forall i \in S_{cus} = \{1, ..., 8\}$) and 3 factories (with their limit prices c^j_t $\forall t \in T = \{9, .., 16\}$, $\forall j \in S_{fac} = \{1, ..., 3\}$ for each time-slot) with the optimal solution (allocated customer for each time-slot). The maximum profit extracted here is $35.

Having formally described the problem, we now reconsider the original contex context of AR&O scheduling discussed earlier on, and note that within this domain, Stranjak *et al.* have previously provided a greedy, non-optimal solution to a similar scheduling problem across multiple factories [8]. As with the ILOG CPLEX solution described above, their solution is centralised in nature and, thus, its computationally complexity increasing exponentially, making the solution intractable for large problems. In the next section, we propose a market-based approach to solving such a problem in a decentralised manner with a linearly increasing computational complexity. We evaluate the *efficiency* of such a mechanism as the ratio of profit of all stakeholders extracted in the mechanism to the profit extracted in the optimal allocation given in Equation 3.

3 The Market-Based Solution

We now detail our market-based solution to the multi-factory scheduling problem detailed in Section 2. Specifically, in this section, we describe our approach; an auction mechanism that allows self-interested, profit-motivated buyers (bidding for customers) and sellers (bidding for factories) to compete for time-slots. The scheduling is then determined by transactions (when a set of bids match with a set of asks) among the buyers and sellers which allocate time-slots to jobs subject to the constraints outlined in Section 2 (and specifically those described by Equation 3). We now describe our market protocol that determines how the agents strategically interact in the market.

3.1 The Market Protocol

The protocol we have developed is a variant of the CDA [9,3], designed to maximise profits in the system. In particular, trading agents are allowed to submit multi-unit bids

ASK ORDERBOOK		
ID	Unit Price	Free Time Slots (9:00 to 16:00)
Seller1	$4.25	[1, 1, 1, 1, 0, 0, 0, 0]
Seller2	$4.5	[1, 1, 0, 0, 0, 0, 0, 0]
Seller1	$5.25	[0, 0, 0, 0, 1, 1, 1, 1]
Seller3	$6	[1, 1, 1, 1, 0, 0, 0, 0]
Seller3	$6	[0, 0, 0, 0, 1, 1, 1, 1]
Seller2	$6.25	[0, 0, 0, 0, 0, 0, 1, 1]
Seller2	$6.75	[0, 0, 1, 1, 1, 1, 0, 0]
...	...	...

BID ORDERBOOK		
ID	Total Price	Quantity	Deadline
Buyer8	$12	3	16
Buyer1	$14.5	4	12
Buyer3	$9.5	3	11
...	...	...	...

(a)   (b)

Fig. 2. (a) Orderbook with uncleared asks, first ordered by lowest bid price per unit and, second, by the earliest time-slots for similar prices. (b) Uncleared bids are queued in an orderbook, ordered by the highest bid price per unit.

and asks (i.e. offers to buy and sell a number of time-slots respectively) which are queued in a bid orderbook (see Figure 2(a)) and an ask orderbook (see Figure 2(b)) respectively. These offers indicate a commitment from the buyers and sellers and cannot be withdrawn. The multi-unit bids allows allow the customers to express the number of time-slots required, and the order books effectively provide the mechanism by which any matching bids and asks are cleared. In more detail, the protocol proceeds as follows:

1. **Bid:** Buyer i submits a multi-unit bid, $bid_i = (id^i_{cus}, p^i_{total}, q^i, t^i_{deadline})$, $t^i_{deadline} \in T$ to buy exactly q time-slots (given the inelastic demand) within its deadline $t_{deadline}$ for no more than a *total price* of p.

2. **Ask:** Seller j submits different multi-unit asks, $ask^a_j = (id^j_{fac}, p^{j,a}_{unit}, f^{j,a}_{T_{start}}, ..., f^{j,a}_{T_{end}})$, $\forall t \in T$, $a \in A^j_{offered} \subset A^j$ for (not necessarily all) the unscheduled time-slots in its factory, with $A^j_{offered}$ defining the set of multi-unit asks. Note that the ask is defined over all the different time-slots to allow multi-unit asks in the market, rather than single-unit asks over single unscheduled time-slots. This is to simplify the bidding process by grouping similar asks.

3. **Bid orderbook:** Bids are queued in a bid orderbook, ordered by the highest price per unit[7] (see Figure 2(a)). Bids cannot be retracted once queued in the order book. This is to ensure consistency in the orderbook such that a seller may accept a bid without the risk of that bid being retracted. Thus, a bid may only be replaced by improving the bid (i.e. submitting a higher price) which would allow buyers to compete by shading their bids.

4. **Ask orderbook:** Asks are queued in an ask orderbook, ordered first by the lowest unit price and second (given the same unit price) by the earliest time-slot (see Figure 2(b)). When we have asks with similar prices, our protocol prioritises the clearing

---

[7] Highest unit-price ordering is necessary because a job's value is defined for the whole job, rather than over the different time-slots required.

of the earlier ones. This is because the later time-slots have a higher probability of clearing future bids and, hence, a higher expected profit in the future than the earlier ones. Thus, our mechanism clears earlier asks with the same price first to maximise profit. As with buyers, sellers are not allowed to retract asks, but are allowed to improve on them by submitting a lower ask price.

5. *Clearing a new bid:* Whenever a new bid is added in the bid orderbook, the market attempts to clear by matching the new bid[8] with the ask orderbook. Our mechanism searches the ask orderbook for the set of lowest asks, $A^j_{matched} \subset A^j$ from each seller j that would completely clear the bid. The market then clears the matched asks $A^{j^*}_{matched}$ from the seller j^*, if any, with the lowest total price against the new bid. If the market clears, the newly matched bid is removed from the bid orderbook while the parameters of the matched asks, $f^{j^*}_t \; \forall t \in T$, are updated. If $\sum_{t \in T} f^{j^*,a}_t = 0$, *where* $a \in A^{j^*}_{matched}$, $ask^a_{j^*}$ is removed from the ask orderbook (because the ask has been completely cleared and all time-slots have been scheduled).

6. *Clearing a new ask:* When a new ask is received, the market attempts to match the seller (with now a better set of asks queued in the ask orderbook) which submitted that ask with the bid orderbook. In particular, the mechanism runs down the bid orderbook to find the highest bids that would be completely cleared by the seller's set of asks (stopping when all the asks from that seller are cleared or at the end of the bid orderbook). The market then clears these highest bids, if any, against the matched asks from the seller. All the cleared bids are then removed from the bid orderbook, while the cleared asks are updated and removed from the orderbook if completely allocated (as seen with the clearing of a new bid).

Given the structure of our market mechanism, we now consider its behaviour. In particular, we use a simple bidding strategy for such a market protocol in order to provide a lower bound benchmark on the efficiency of the market mechanism.

3.2   The Bidding Strategy

One of the principal concerns in developing a market mechanism is to ensure that it is efficient and that the system does not break down even with comparatively simple bidding behaviour on behalf of the buyers and sellers. This is important because as designers, we cannot dictate the specific strategies of the buyers and sellers and, so, we want to ensure that the market performs well for whatever strategies are adopted. The underlying intuition here is that by considering this simple behaviour, we are able to establish a lower bound on the efficiency credited principally to the market structure rather than the behaviour (assuming that agents are motivated by profits and not malicious, e.g. sellers bidding less and buyers more than their limit price to break down the market). This approach has been advocated a number of researchers, most notably by Gode and Sunder [5], and thus, to this end, we adopt Gode and Sunder's Zero-Intelligence (ZI)

---

[8] Our mechanism attempts to clear only the new bid (ask) because we are continuously clearing the market which ensures that any queued bid (ask) cannot be cleared by the current ask (bid) orderbook.

bidding strategy in our work because it simply submits a random bid or ask based solely on its limit price, ignoring the state of the market or past market information [4]. The Zero-Intelligence strategy works as follows[9]:

1. For buyer $i \in S_{cus}$,

$$p^i \sim \mathcal{U}(0, \ell_i)$$
$$\text{bid}_i = (id^i_{cus}, p^i, q^i, t^i_{deadline})$$

2. For seller $j \in S_{fac}$,

$$p^{j,a} \sim \mathcal{U}(c^a_i, p_{max})$$
$$\text{ask}^a_j = (id^j_{fac}, p^{j,a}, f^{j,a}_{T_{start}}, ..., f^{j,a}_{T_{end}}) \ \forall a \in A^j$$

Thus, the buyer submits a multi-unit bid based on its limit price at random times. Conversely, the seller j submits a set of multi-unit asks, A^j, over all the unscheduled time-slots in its factory also at random times. Given our market protocol and the Zero-Intelligence strategy, we now empirically evaluate our market-based scheduling mechanism.

4   Empirical Analysis

In this section, we empirically compare the market-based solution (see Section 3) against the optimal solution (see Section 2). In our experimental setup, for every combination of buyers and sellers, we consider 50 different sets of demand and supply (see Equations 1 and 2). For each set of demand and supply, we consider a statistically significant number of runs[10], namely 100, and average the performance over these different runs and sets of demand and supply. Based on standard experimental setup of the CDA [9], we induce the demand and supply by drawing buyer i's and seller j's endowment (of time-slots) from random distributions[11] as follows:

$$T_{start} = 9$$
$$T_{end} = 16$$
$$q^i \sim \mathcal{U}^\mathcal{I}(1, 4),$$
$$t^i_{deadline} \sim \mathcal{U}^\mathcal{I}(q^i - 1, (T_{end} - T_{start})) + T_{start},$$
$$\ell^i \sim \mathcal{U}(1.5, 4.5) \times t^i_{deadline}$$
$$c^{j,a} \sim \mathcal{U}(1.5, 4.5), \ \forall a \in A^j$$

where A^j is a randomly generated set of sets grouping similar limit prices. Furthermore, because of the informationally decentralised nature of the mechanism, it is not possible to determine when the market reaches completion[12]. Thus, we impose a deadline to

---

[9] $X \sim \mathcal{U}(a, b)$ is a discrete uniform distribution between a and b with steps of 0.01.

[10] We validated our results at the 95%-confidence interval by running the non-parametric Wilcoxon rank-sum test.

[11] $X \sim \mathcal{U}^\mathcal{I}(a, b)$ is a uniform distribution of integers between a and b.

[12] A market reaches completion when there can be no more transactions. This information is unknown unless all private information is available.

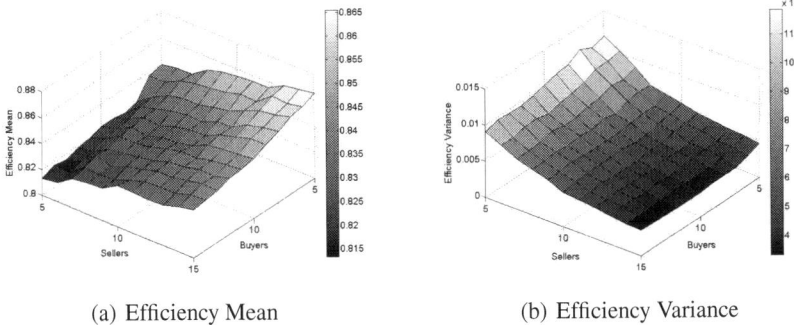

(a) Efficiency Mean   (b) Efficiency Variance

Fig. 3. The efficiency of the market mechanism for different numbers of buyers and sellers in the market. Efficiency converges to an efficiency of 84% as the number of buyers and sellers increases, while the variance decreases.

limit the duration of the auction[13]. In our experiments, we set this deadline to 5000 rounds.

The mean and variance of the efficiency (defined in Section 2) of the market-based solution over different problem sizes is given in Figures 3(a) and 3(b) respectively. As we can observe, the market efficiency averages 84% (ranging between 81% and 87%) with efficiencies converging to 84% as the number of factories and consumers increases. We also observe that the variances of the efficiency decreases rapidly as the size of the scheduling problem increases. Thus, our mechanism becomes more effective in finding profitable allocations as the number of factories increases while its efficiency is unaffected by increasing demand.

Now, because we impose a deadline (as it is unknown as to when no more resources can be cleared in the market), it is insightful to analyse if we are closing the market too early or too late. To this effect, in an example of a large problem with 15 factories and 15 customers[14], we consider how the efficiency and the volume of allocation of time-slots vary over the rounds (see Figure 4). In particular, we observe that the bulk of the allocations are made within the first few hundreds rounds (with 85% allocated within the first 500 rounds) even though the market reaches completion after 3000 rounds. This validates our choice for a deadline at 5000 rounds as we effectively limit the duration of the auction without compromising on efficiency. Furthermore, because time-slots are gradually allocated, we can consider our market-based mechanism as an *any-time* approach (which can be halted at any time for a solution). This contrasts with the centralised approach where time-slots are only allocated once a solution is computed. An any-time solution would indeed be very useful in a problems with hard deadlines.

Furthermore, to examine the tractability of our market-based solution, we compare the computational time of our market mechanism against that of a centralised, optimal solution computed using ILOG CPLEX (as highlighted in Section 2). While the

---

[13] If we consider an environment where agents are allowed to enter or leave the system or can renew their endowment, we do not impose a deadline in our auction which possibly never runs out of transactions.

[14] Similar trends were observed for other numbers of factories and customers.

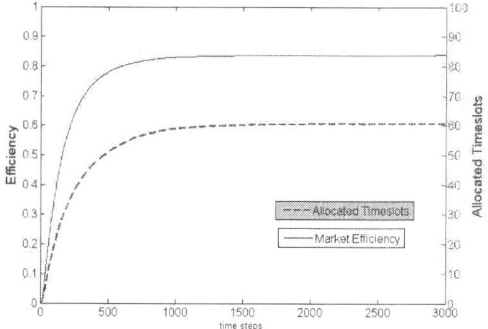

Fig. 4. For a problem of 15 factories and 15 customers, the market closes after 3000 rounds at an efficiency of 84%. Note that 84% of the volume of transactions are completed within the first 500 rounds and 97% within 1000 rounds. The efficiency after 500 rounds is within 7% of its maximum and within 0.9% after 1000 rounds.

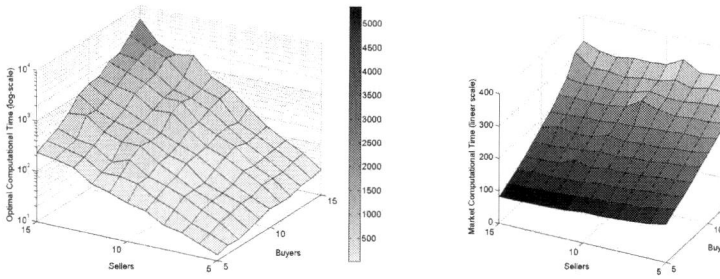

(a) Computational time for optimal solution (log-scale)

(b) Computational time for market-based solution

Fig. 5. Computational times of the centralised (a) and decentralised (b) approaches. Note that the former increases exponentially while the latter scales linearly. For a problem with 15 factories and 15 customers, the computational time for the optimal solution is around 5400ms compared to 320ms for the market-based one.

computational time of the centralised, optimal solution increases exponentially, that of the decentralised market-based approach increases linearly. This is because the computational complexity of the latter approach is due principally to the clearing process, with the size of the orderbooks to be cleared also increasing linearly. Thus, our market-based approach is indeed tractable as one of the desirable properties for a decentralised mechanism in the motivating AR&O setting.

5   Conclusions and Future Work

In this paper, we proposed a novel decentralised mechanism for multi-factory scheduling based on a variant of the Continuous Double Auction, that does not require the revelation of private preferences to a third-party agent. We empirically demonstrated a lower bound efficiency of 84% in our auction mechanism using a Zero-Intelligence bidding strategy. We thus showed that we sacrifice a reasonably small level of efficiency for the benefits of a decentralised and transparent approach (through its public order-books and the fact that there is no center) and scaleability (given the linearly increasing complexity of our market-based solution).

Our future work in this area focuses on taking the solution that we have developed here and applying it within a fine grained industrial simulation of the Aero Repair and Overhaul domain (before ultimately deploying it within the operational system). As we envisage increasingly larger scheduling problems in this context, the need for robust and scaleable solutions of the kind that we have presented here, will become highly desirable and indeed essential in the future. To further improve the performance of our scheduling approach we would also like to explore more intelligent bidding strategies that can strategise effectively on the additional time factor that we have in this domain. In this respect, we believe that a considerably higher efficiency can be achieved, and this belief is supported by the observation that within the standard CDA, state of the art strategies can led to an improvement in efficiency from 97% (achieved with the Zero-Intelligence bidding strategy) to 99.9% [11,10]. In addition, we intend to analyse the efficiency of our system when agents enter and leave the system at any time. This is an important issue within the Aero Repair and Overhaul domain, since damaged engines must often be added to an existing schedule at short notice, and thus, we must evaluate how well our auction-based mechanism reacts to sudden change in the demand and supply.

References

1. Baker, A.D.: Metaphor or reality: A case study where agents bid with actual costs to schedule a factory. In: Market-based Control, ed. Clearwater. World Scientific, New York (1992)
2. Blazewicz, J., Ecker, K.H., Pesch, E., Weglarz, J.: Scheduling Computer and Manufacturing Processes. Springer, Berlin (1996)
3. Friedman, D., Rust, J.: The Double Auction Market: Institutions, Theories and Evidence. Addison-Wesley, New York (1992)
4. Gode, D.K., Sunder, S.: Allocative efficiency of markets with zero-intelligence traders: Market as a partial substitute for individual rationality. Journal of Political Economy 101(1), 119–137 (1993)
5. Gode, D.K., Sunder, S.: What Makes Markets Allocationally Efficient? The Quarterly Journal of Economics 35, 603–630 (1997)
6. Ichimi, N., Iima, H., Hara, T., Sannomiya, N.: Autonomous decentralized scheduling algorithm for a job-shop scheduling problem with complicated constraints. In: Proceedings the Fourth International Symposium on Autonomous Decentralized Systems, pp. 366–369 (1999)
7. Rassenti, S.J., Smith, V.L., Bulfin, R.L.: A combinatorial auction mechanism for airport time slot allocation. The Bell Journal of Economics 13(2), 402–417 (1982)
8. Stranjak, A., Dutta, P.S., Ebden, M., Rogers, A., Vytelingum, P.: A multi-agent simulation system for prediction and scheduling of aero engine overhaul. In: Proceedings of the seventh International Conference on Autonomous Agents and Multi-Agent Systems (Industrial Track), pp. 81–88 (2008)

9. Vytelingum, P.: The structure and behaviour of the Continuous Double Auction, Ph.D. dissertation, School of Electronics and Computer Science, University of Southampton (2006)
10. Vytelingum, P., Cliff, D., Jennings, N.R.: Strategic bidding in continuous double auctions. Artificial Intelligence Journal 172(14), 1700–1729 (2008)
11. Vytelingum, P., Dash, R.K., David, E., Jennings, N.R.: A risk-based bidding strategy for continuous double auctions. In: Proceedings of the Sixteenth European Conference on Artificial Intelligence, pp. 79–83 (2004)
12. Wellman, M., Walsh, W., Wurman, P., MacKie-Mason, J.: Auction protocols for decentralized scheduling. Games and Economic Behavior 35, 271–303 (2001)

Auctions with Dynamic Populations: Efficiency and Revenue Maximization
(Extended Abstract)*

Maher Said

Department of Economics
Yale University
New Haven, CT, USA
maher.said@yale.edu

Abstract. We study a stochastic sequential allocation problem with a dynamic population of privately-informed buyers. We characterize the set of efficient allocation rules and show that a dynamic VCG mechanism is both efficient and periodic ex post incentive compatible; we also show that the revenue-maximizing direct mechanism is a pivot mechanism with a reserve price. We then consider sequential ascending auctions in this setting, both with and without a reserve price. We construct equilibrium bidding strategies in this indirect mechanism where bidders reveal their private information in *every* period, yielding the same outcomes as the direct mechanisms. Thus, the sequential ascending auction is a natural institution for achieving either efficient or optimal outcomes.

Keywords: Dynamic mechanism design, Random arrivals, Dynamic Vickrey-Clarke-Groves, Sequential ascending auctions.

In this paper, we study the problem of a seller faced with a stochastic sequential allocation problem. At each point in time, a random number of buyers and objects arrive to a market. Buyers are risk-neutral and patient, while objects are homogeneous and perishable. Moreover, each buyer desires a single unit of the good in question; however, valuations for the good vary across buyers. The mechanism designer must elicit the private information of these buyers in order to achieve a desirable allocation—one that is either efficient or revenue maximizing, depending on the designer's objective function.

We are concerned with two main questions. First, what outcomes are attainable in markets with dynamic populations of privately-informed buyers? In particular, can we achieve efficient or revenue-maximizing outcomes? And secondly, and equally importantly, can we achieve these outcomes using natural or simple "real-world" institutions?

It is well-known that the Vickrey-Clarke-Groves mechanism is efficient and dominant-strategy incentive compatible in static environments. By choosing a price for each buyer that equals the externality imposed by her report on other participants in the mechanism, the VCG mechanism aligns the incentives of

---

* The complete paper is available online at http://ssrn.com/abstract=1296455

S. Das et al. (Eds.): Amma 2009, LNICST 14, pp. 87–88, 2009.

the buyer with those of a welfare-maximizing social planner. In the dynamic environment we consider, the arrival of a new buyer imposes an externality on her competitors by reordering the (anticipated) schedule of allocations to those buyers currently present on the market, as well as to those buyers expected to arrive in the future. We show that by charging each agent, upon her arrival, a price equal to this *expected* externality, the buyer's incentives are aligned with those of the forward-looking planner. Therefore, this dynamic version of the VCG mechanism is efficient and periodic ex post incentive compatible.

This dynamic VCG mechanism is a direct revelation mechanism, requiring buyers to report their values to the mechanism upon their arrival to the market. As is well known, however, direct revelation mechanisms may be difficult to implement in practice. We therefore turn to the design of simple indirect mechanisms and consider the possibility of achieving efficient outcomes via a sequence of auctions. Despite the resemblance of the dynamic direct mechanism to its single-unit static counterpart, we find that this relationship does not hold for the corresponding auction format. In the canonical static auction environment, the "standard" analogue of the VCG mechanism is the second-price sealed-bid auction. In our dynamic environment, however, a sequence of such auctions cannot yield outcomes equivalent to those of the dynamic VCG mechanism. In a sequential auction, there is an "option value" associated with losing in a particular period, as buyers can participate in future auctions. The value of this option depends on the private information of all other competitors, as the expected price in the future will depend on their values—despite the assumption of independent private values, the market dynamics generate interdependence.

Therefore, a standard second-price sealed-bid auction does not reveal sufficient information for the correct determination of buyers' option values. In contrast, the ascending auction is an *open* auction format that allows for the gradual revelation of private information. We use this to construct bidding strategies that form an efficient equilibrium in a sequence of ascending auctions. In each period, buyers bid up to the price at which they are indifferent between winning an object and receiving their expected marginal contribution to the social welfare in the future. As buyers drop out of the auction, they reveal their private information to their competitors, who then condition their current-period bids on this information. This process of information revelation is repeated in *every* period, and is crucial for providing the appropriate incentives for new entrants to also reveal their private information. We show that these memoryless bidding strategies form a periodic ex post equilibrium that is outcome equivalent to the truth-telling equilibrium of the efficient direct mechanism.

In addition, we construct a revenue-maximizing direct mechanism for this setting. We show that the revenue-maximizing direct mechanism is the dynamic VCG mechanism applied to buyers' *virtual* values and the *virtual* surplus. When buyers' values are drawn from the same distribution, the sequential ascending auction with an optimal reserve price admits an equilibrium that is equivalent to truth-telling in the optimal direct mechanism. Thus, the sequential ascending auction is a natural institution for achieving either efficient or optimal outcomes.

Revenue Submodularity*

Shaddin Dughmi**, Tim Roughgarden* * *, and Mukund Sundararajan[†]

Department of Computer Science, Stanford University, Stanford CA 94305, USA
{shaddin,tim,mukunds}@cs.stanford.edu

Abstract. We introduce *revenue submodularity*, the property that market expansion has diminishing returns on an auction's expected revenue. We prove that revenue submodularity is generally possible only in matroid markets, that Bayesian-optimal auctions are always revenue-submodular in such markets, and that the VCG mechanism is revenue-submodular in matroid markets with IID bidders and "sufficient competition". We also give two applications of revenue submodularity: good approximation algorithms for novel market expansion problems, and approximate revenue guarantees for the VCG mechanism with IID bidders.

1 Revenue Submodularity

Auction environments are never static. Existing bidders may leave and new bidders may join or be recruited. For this reason, it is important to study how auction revenue *changes* as a function of the bidder set.

We introduce *revenue submodularity* — essentially, the property that market expansion has diminishing returns on an auction's expected revenue. For example, in a multi-unit auction with bidders that have unit demand and IID valuations, revenue submodularity means that the auction's expected revenue is a concave function of the number of bidders. In general, an auction A is deemed revenue submodular in an environment with potential bidders U if, for every subset $S \subset U$ and bidder $i \notin S$, the increase in the auction's revenue from supplementing the bidders S by i is at most that of supplementing a set $T \subseteq S$ of bidders by the same additional bidder i.

We first identify the largest class of single-parameter domains for which general revenue-submodular results are possible: *matroid markets*, in which the feasible subsets of simultaneously winning bidders form a matroid. Fortunately, matroid markets include several interesting examples, including multi-unit auctions and unit-demand matching markets (corresponding to a transversal matroid).

---

* A working paper featuring special cases of results in this work was presented at the 3rd Workshop on Sponsored Search, May 2007, under the title "Is Efficiency Expensive?".
** Supported in part by NSF grant CCF-0448664.
* * * Supported in part by NSF CAREER Award CCF-0448664, an ONR Young Investigator Award, an AFOSR MURI grant, and an Alfred P. Sloan Fellowship.
† Supported by NSF Award CCF-0448664 and a Stanford Graduate Fellowship.

S. Das et al. (Eds.): Amma 2009, LNICST 14, pp. 89–91, 2009.
© ICST Institute for Computer Sciences, Social-Informatics and Telecommunications Engineering 2009

We then prove a number of positive results. First is a sweeping result for (Bayesian)-optimal auctions: in every matroid market with independent (not necessarily identical) valuation distributions, the revenue-maximizing auction is revenue-submodular. The VCG mechanism, on the other hand, enjoys revenue-submodularity only under additional conditions, even when valuations are IID draws from a well-behaved distribution. We identify the key sufficient condition under which the VCG mechanism is revenue-submodular with IID bidders, which is a matroid rank condition stating that there is "sufficient competition" in the market. For example, in multi-unit auctions (uniform matroids), sufficient competition requires that the number of bidders is at least the number of items. Finally, we prove that revenue-submodularity is not a monotone property of the reserve prices used: reserve prices higher than those in an optimal mechanism preserve submodularity , but reserve prices strictly between those in the VCG mechanism (zero) and those in an optimal mechanism always have the potential to destroy revenue-submodularity, even when there is sufficient competition in the market. Formally, our main results are as follows.

Theorem 1 (Submodularity of Optimal Auctions in Matroid Markets)
Fix a matroid market M over bidders U with valuations drawn independently from arbitrary distributions $\{F_i\}_{i \in U}$. The expected revenue of the corresponding optimal auction for induced matroid markets M_S is submodular over $S \subseteq U$.

Theorem 2 (Submodularity of the VCG Mechanism on Full-Rank Sets)
Fix a matroid market M over bidders U with valuations drawn IID from a regular distribution F. The expected revenue of VCG for induced matroid markets M_S is submodular over the full-rank sets $S \subseteq U$.

Theorem 3 (Submodularity with Incorrect Reserve Prices)

(a)   *For every regular distribution F with optimal reserve price r^*, every matroid market with bidders U with valuations drawn IID from F, and every $r \geq r^*$, the expected revenue of the VCG mechanism with reserve price r is submodular on U.*

(b)   *For every $\epsilon \in (0,1)$, there is a regular distribution F with optimal reserve price r^* and a matroid market for which the expected revenue of the VCG mechanism with reserve price $(1 - \epsilon)r^*$ is not submodular on full-rank sets.*

We obtain reasonably simple and direct proofs of these results by appropriately applying, in different ways, two elegant but powerful techniques: Myerson's characterization [2] of expected auction revenue in terms of the expected "virtual surplus" of the auction's allocation; and the submodularity that arises from optimizing a nonnegative weight vector over the independent sets of a matroid.

2   Applications

The first application of revenue submodularity is algorithmic. In the basic version of the *market expansion problem*, we are given a matroid market with a set of potential bidders, a subset of initial bidders, an auction (defined for all induced submarkets), and an expansion budget k. The goal is to recruit a set of at most k new bidders to maximize the expected revenue of the auction on the submarket induced by the original bidders together with the new recruits. This problem is easy only when the environment is completely symmetric (IID bidders in a multi-unit auction). Using the result of Nehmauser, Wolsey, and Fisher [3], we observe that "greedy market expansion" — repeatedly adding the new bidder that (myopically) increases the expected revenue of the auction as much as possible — is a constant-factor approximation algorithm provided the given auction is revenue submodular over all sets containing the initial bidders.

Our second application of revenue submodularity is to approximate revenue-maximization guarantees for the VCG mechanism. More precisely, in a matroid market with IID bidder valuations and "modest competition" — formalized using matroid connectivity — the VCG mechanism always obtains a constant fraction of the revenue of an optimal auction; moreover, the approximation guarantee tends rapidly to 1 as the degree of competition increases. As part of our proof, we generalize to arbitrary matroids the bicriteria result of Bulow and Klemperer [1]. This result suggests an explanation for the persistent use of efficient auctions in settings where the auctioneer should presumably care about maximizing revenue. In many contexts, the cost (i.e., revenue loss) of running an efficient auction is small and outweighed by the benefits (relative simplicity and optimal efficiency) even for a revenue-maximizing seller. We also extend this result to a standard model of sponsored search auctions.

References

1. Bulow, J., Klemperer, P.: Auctions versus negotiations. American Economic Review 86(1), 180–194 (1996)
2. Myerson, R.: Optimal auction design. Mathematics of Operations Research 6(1), 58–73 (1981)
3. Nemhauser, G., Wolsey, L., Fisher, M.: An analysis of the approximations for maximizing submodular set functions. Mathematical Programming 14, 265–294 (1978)

Fair Package Assignment

Sébastien Lahaie[1] and David C. Parkes[2]

[1] Yahoo Research
New York, NY 10018
lahaies@yahoo-inc.com
[2] School of Engineering and Applied Sciences
Harvard University
Cambridge, MA 02138
parkes@eecs.harvard.edu

Abstract. We consider the problem of fair allocation in the package assignment model, where a set of indivisible items, held by single seller, must be efficiently allocated to agents with quasi-linear utilities. A fair assignment is one that is efficient and envy-free. We consider a model where bidders have superadditive valuations, meaning that items are pure complements. Our central result is that core outcomes are fair and even coalition-fair over this domain, while fair distributions may not even exist for general valuations. Of relevance to auction design, we also establish that the core is equivalent to the set of anonymous-price competitive equilibria, and that superadditive valuations are a maximal domain that guarantees the existence of anonymous-price competitive equilibrium. Our results are analogs of core equivalence results for linear prices in the standard assignment model, and for nonlinear, non-anonymous prices in the package assignment model with general valuations.

S. Das et al. (Eds.): Amma 2009, LNICST 14, p. 92, 2009.
© ICST Institute for Computer Sciences, Social-Informatics and Telecommunications Engineering 2009

Solving Winner Determination Problems for Auctions with Economies of Scope and Scale

Martin Bichler[1], Stefan Schneider[1], Kemal Guler[2], and Mehmet Sayal[2]

[1] Technische Universität München
Boltzmannstrasse 3, 85748 Garching, Germany
`{bichler,schneist}@in.tum.de`
[2] HP Laboratories
1501 Page Mill Rd., Palo Alto, CA 94304, USA
`{kemal.guler,mehmet.sayal}@hp.com`

Abstract. Economies of scale and scope describe key characteristics of production cost functions that influence allocations and prices on procurement markets. Auction designs for markets with economies of scale are much less well understood than combinatorial auctions, they require new bidding languages, and the supplier selection typically becomes a hard computational problem. We suggest a bidding language for respective markets, and conduct computational experiments to explore the incremental computational burden to determine optimal solutions brought about by the need to express economies of scope for problems of practical size.

Keywords: volume discount auctions, procurement auctions, economies of scale, economies of scope.

1 Introduction

This paper is motivated by real world procurement practices of an industry partner, where procurement managers need to purchase large volumes of multiple items. Typically, in these circumstances, economies of scale are jointly present with economies of scope. For example, suppliers that set up a finishing line for a certain product have high setup costs, but low marginal cost leading to a unit price degression. Economies of scope often arise in shipping and handling a larger number of items to a customer.

State of the practice in markets with economies of scale are split-award contracts, where the best bidder gets the larger share of the volume for a particular quantity and the second best bidder gets a smaller share (e.g., a 70/30% split). With significant economies of scale, suppliers face a strategic problem in these auctions, since there is uncertainty about which quantity they will get awarded, they might speculate and bid less aggressive based on the unit cost for the smaller share. In other words, split award auctions do not allow suppliers do adequately express economies of scale.

2 Bid Language

Our bid language describes the types of bids that can be submitted and is motivated by real-world business practice. We suggest tiered bids, a bid language where suppliers

S. Das et al. (Eds.): Amma 2009, LNICST 14, pp. 93–94, 2009.
© ICST Institute for Computer Sciences, Social-Informatics and Telecommunications Engineering 2009

submit unit prices that are valid for continuous intervals of the volume of items purchased. In contrast to Davenport and Kalagnanam [1] the quantity purchased determines a unit price that is valid for the entire quantity.

In addition, the suppliers can indicate rebates of the form: "If you buy a quantity of more than x units of various items you will receive an additional rebate of y". This allows bidders to express economies of scope that span more than one item. Additionally bidders often specify these kinds of rebates out of strategic considerations, in order to secure a share of the market.

3 Computational Complexity of the Supplier Selection

While the bid language leads to much flexibility for the suppliers in specifying discounts and rebates in a compact format, it also incurs computational complexity on the buyer's side. We show that the problem is weakly NP-complete and explore the empirical hardness of the supplier selection problem if solved exactly with a branch-and-bound approach. The supplier selection process should be used in an interactive manner exploring different award scenarios and side constraints. The process is sometimes refered to as scenario analysis. Therefore, the experiments focused on solver times of up to an hour.

The results are based on a large set of experiments with real and synthetic bid data. The synthetic bid data follows the characteristics that we found in real-world bid data. We observed that, with tiered bids but without rebates, the optimal solution was found in a matter of minutes for problem sizes of up to 12 suppliers, 24 items, 4 tiers, and 5 rebate schedules. Adding 8 discount tiers will often lead to solution times of several hours. The results provide an understanding which problem sizes can be solved optimally in an interactive decision support tool.

Reference

1. Davenport, A., Kalagnanam, J.: Price negotiations for procurement of direct inputs. In: Dietrich, B., Vohra, R. (eds.) Mathematics of the Internet: E-Auction and Markets. IMA Volumes in Mathematics and its Applications. Springer, Heidelberg (2001)

Running Out of Numbers: Scarcity of IP Addresses and What to Do about It

Benjamin Edelman

Harvard Business School
bedelman@hbs.edu

Abstract. The Internet's current numbering system is nearing exhaustion: Existing protocols allow only a finite set of computer numbers ("IP addresses"), and central authorities will soon deplete their supply. I evaluate a series of possible responses to this shortage: Sharing addresses impedes new Internet applications and does not seem to be scalable. A new numbering system ("IPv6") offers greater capacity, but network incentives impede transition. Paid transfers of IP addresses would better allocate resources to those who need them most, but unrestricted transfers might threaten the Internet's routing system. I suggest policies to facilitate an IP address "market" while avoiding major negative externalities – mitigating the worst effects of v4 scarcity, while obtaining price discovery and allocative efficiency benefits of market transactions.

Keywords: Market design, IP addresses, network, Internet.

Disclosure. I advise ARIN's counsel on matters pertaining to v4 exhaustion, v6 transition, and possible revisions to ARIN policy. This paper expresses only my own views – not the views of ARIN or of those who kindly discussed these matters with me.

1 Introduction

Hidden from view of typical users, every Internet communication relies on an underlying system of numbers to identify data sources and destinations. Users typically specify online destinations by entering domain names (*e.g.* "congress.gov"). But the Internet's routers forward data according to numeric IP addresses (*e.g.* 140.147.249.9).

To date, the Internet has enjoyed an ample supply of IP addresses. The Internet's standard "IPv4" protocol offers 2^{32} addresses ($\approx$4.3 billion). But demand is substantial and growing. Current allocation rates suggest exhaustion by approximately 2011 [1].

Engineers have developed a new numbering system, *IPv6*, which offers 2^{128} possible addresses (more than three billion billion billion). But incentives hinder transition, as detailed in Section 0. The Internet therefore faces the prospect of continuing to rely on the current IPv4 address system even after v4 addresses "run out." v4 scarcity will limit future expansion, hinder some technologies, and impose new costs on networks and users.

S. Das et al. (Eds.): Amma 2009, LNICST 14, pp. 95–106, 2009.

This paper proceeds as follows: In Section 0, I present the technology of IP addressing, and the institutions and policies that allocate addresses. I then turn to specific tactics to manage scarcity. In Section 0, I evaluate central planning, and in Section 0 I examine address sharing. In Section 0, I consider IPv6, including factors impeding transition. In Section 0, I explore a market mechanism to reallocate v4 addresses through transfers; I assess key externalities and policy responses.

2   The Technology and Institutions of IP Addressing

IP addresses were first distributed by computer scientists at the Information Sciences Institute (ISI). Initially, scarcity seemed unlikely: Computers were costly, few networks wanted Internet connections, and IPv4 offered billions of addresses. But in the interest of good stewardship, ISI attempted to grant address blocks matching networks' needs. The US military, defense contractors, and large universities received "Class A" blocks (2^{24} addresses, approximately 16.7 million). "Class B" (2^{16}) and "C" blocks (2^8) were provided to smaller networks. Early network architecture permitted only these three sizes.

As demand grew, address assignment developed a geographic hierarchy. The Internet Assigned Numbers Authority (IANA) now grants large "/8" (read: "slash eight") blocks of 2^{24} addresses to Regional Internet Registries (RIRs). RIRs in turn assign addresses within their regions. Initial RIRs were RIPE NCC (for Europe, the Middle East, and parts of Africa), APNIC (for the Asia-Pacific region), and ARIN (North America and, at the time, Latin America and parts of Africa). Later, RIRs opened in Africa and Latin America.

RIRs seek to satisfy network operators' demonstrated address needs. An interested network submits a request for addresses to its RIR, along with documents showing its need and its exhaustion of any previously-granted addresses. (Documentation often includes equipment receipts, customer lists, or business plans.) RIR fees follow the principle of cost recovery, rather than maximizing RIR revenue or profit. For example, the largest US networks pay ARIN just $18,000 per year.

IANA continues to assign addresses to RIRs. But IANA's *free pool* reveals an impending shortage: As of January 2009, only 34 /8's remain, and RIRs have recently claimed 6 to 12 /8's per year [1]. Even if demand does not accelerate as exhaustion nears, it seems IANA will soon have no more addresses left to provide to RIRs. Projections for IANA's *v4 free pool exhaustion* range from June 2010 [2] to March 2011 [1].

3   Relieving v4 Scarcity through Central Planning

In principle, central authorities could ease IPv4 scarcity by requiring that networks, *e.g.*, migrate to IPv6 as presented in Section 0, on pain of losing ongoing RIR services. Networks want to be listed in RIR *Whois* records so that others can confirm their rights in the corresponding addresses, and networks want RIR *reverse addressing* services so that automated systems can confirm the host names associated with a network's addresses. These services will be increasingly valuable if IP addresses come to be viewed as scarce resources to be safeguarded and potentially exchanged for value.

But in practice, central authorities have limited power to force migration to v6. Once a network receives addresses from an RIR, it does not directly need substantial ongoing RIR service. Whois primarily benefits the larger community by telling others how to reach the network's technical contacts. Thus, withholding Whois would little threaten an existing network. Even if an RIR declared that a network could no longer use its existing addresses, other networks would continue to know the target network by those addresses, so the network could keep the block with impunity. Whois records are more important when a network seeks to change its connectivity, for an ISP typically checks Whois to confirm a network's rights in the addresses it seeks to use. But if address revocations take effect only upon a connectivity change, most networks could ignore revocations for some time, and networks could retain their existing connectivity to avoid losing addresses. In the future, resource certification might link inter-network communications to RIRs' attestations of address ownership, but such linkages are not yet developed [3].

Institutions and norms also constrain central authorities' ability to force migration. Networks control RIRs through periodic elections of RIR directors, so RIRs cannot act contrary to networks' perceived interests. Furthermore, networks have agreed that RIRs serve principally as custodians to assure that resources are allocated uniquely; networks would oppose RIRs forcing use of particular technologies.

Governments are also badly positioned to accelerate v6 implementation. The Internet's worldwide reach defies control by any single country. Furthermore, even large countries struggle to push transition. For example, the US Office of Management and Budget in 2005 set a June 2008 deadline by which federal agency network backbones must support IPv6 [4] – but compliance devolved into installing equipment that need not actually be used [5]. Japanese tax incentives were similarly ineffective in converting users to v6 [6].

4 Sharing IP Addresses to Reduce v4 Demand

As new IPv4 addresses become scarce, some network operators may seek to share addresses among multiple devices. Consider the *home gateway* many users today connect to their cable or DSL modems, letting multiple devices share a single Internet connection and a single public v4 address. Through *Network Address Translation* (NAT), a gateway "rewrites" each outbound IP packet so that, from the perspective of outside networks, that packet comes from the single v4 address assigned to the gateway. When an inbound packet arrives, the gateway attempts to determine which device should receive that packet.

In principle, ISPs can implement similar address translation on a large scale. An interested ISP would assign its users private addresses, using NAT to consolidate onto fewer public addresses at the border between the ISP and the public Internet. Just as many companies offer "extension 101" on their respective phone exchanges, each private IP address may be used simultaneously by many users around the world.

Despite benefits for address conservation, NAT imposes serious disadvantages. For one, NAT is incompatible with certain communication protocols. In general, it is difficult to send a message to a specific computer when that computer is behind a NAT gateway: The gateway does not know which of its users is the intended recipient of a given inbound message. NAT works well for protocols that begin with a user making a request (*e.g.* requesting a web page): The gateway sees the initial request and can route the response to the appropriate requestor. But consider, *e.g.*, IP-based telephone service. A gateway cannot easily determine which user should receive a given incoming call. Indeed, standard SIP VoIP calls do not work if both caller and callee are behind NAT.

More generally, NAT interferes with the Internet's end-to-end principle [7], limiting future communication designs and impeding development of certain kinds of new applications. Of course existing NAT *already* imposes these impediments, requiring most consumer-focused systems to accommodate NAT in some way. (For example, Skype developed a system of supernode relays to transport data among to and from NAT users.) But increasingly widespread use of NAT would further complicate such designs and further constrain some kinds of innovation. (Indeed, supernode system failure caused Skype's two-day outage in August 2007.) Network architects therefore consider NAT a poor architecture for widespread future use.

5   IPv6: The Solution to v4 Scarcity?

As early as 1990, engineers recognized the possible future shortage of IPv4 space [8]. A new version of the IP specification, ultimately named IPv6, dramatically expands the numeric address space – offering 2^{128} possible addresses. If many networks moved to v6 and ceased to need or use v4, v4 scarcity would disappear.

5.1   Transition to IPv6

Transition to IPv6 is discouraged in part by the limited benefits of v6. v6 was designed to improve authentication, security, and automatic device configuration [9]. However, most enhanced v6 features were "backported" to be available in IPv4 also. For an individual network considering transition, v6 therefore offers little direct benefit.

Transition to IPv6 is further hindered by limited compatibility both forward (existing IPv4 devices seeking to communicate with v6 devices) and backward (v6 devices communicating with v4). Because v4 and v6 use different header formats, direct v4-v6 communications are impossible. For example, a v6-only device cannot directly access the vast majority of the current web because most web servers currently support only v4.

IPv4-v6 translators appear to be practical for specific individual protocols. For example, a dual-stack proxy server could readily accept HTTP requests on an v6 interface, obtain the requested web pages via v4, and forward responses to the requesting users via v6. But seamlessly integrating such a proxy adds considerable complexity: Either v6-only hosts must recognize servers they can only contact via a proxy, or DNS servers must intercept v6-only devices' requests for v4-only servers [10]. Furthermore, some protocols defy translation – for example, by embedding IP addresses within their payloads or by encrypting communications in a way that stymies translation (as in HTTPS). Facing the complexity, unreliability, and unpredictability translation inevitably introduces, the IETF in July 2007 abandoned [12] the official design of a general-purpose translator [11].

For lack of robust translation, some software and protocols may not function on IPv6-only devices. For example, a v6-only PC might use a v6-to-v4 proxy to browse the web – yet be unable to play online games or make voice-over-IP phone calls that work fine for v4 users, because no proxy exists (or is correctly configured) for those protocols. Because a separate proxy must be designed for each application, some applications may never work over v6 – especially old systems and custom software developed for a particular business or industry. Thus, even though Windows Vista and Mac OS X support v6, few users are likely to consider v6-only networking a desirable choice in the short run. In trials at RIR meetings, networking experts found that v6-to-v4 translation worked well for the web, but services as common as HTTPS, Skype and iChat were unavailable [13].

Further constraining IPv6 deployment, few tools are available for administering v6 on large networks. Tools for network management and security are currently largely available only for v4 networks, and some categories of tools lack any effective v6 implementations [14]. In principle, market forces could encourage the provision of v6 tools. But with most networks currently operating only v4, developers see a limited market for v6 versions – providing little immediate incentive for developing v6 tools.

5.2 Individual Incentives in IPv6 Transition

Transition to IPv6 is hindered by the incentives of individual participants. Consider a network evaluating v6 to reduce its need for v4 addresses. Little web content is available via v6, nor are other important Internet resources available directly to v6-only devices. The network could use v4-v6 translation, but translation adds complexity – inevitable extra costs when applications do not work as expected. Meanwhile, for lack of v6 administration tools, network administrators find v6 more costly and less flexible than v4. The network's deployment of v6 is further stymied by the lack of v6 *transit*: Most ISPs do not provide v6 connectivity [15]. Furthermore, ISPs that provide v6 tend to offer it less reliably than v4, *i.e.* without service level agreements [16,17], with lower reliability, and with greater latency [15]. In short, under current conditions, v6 is an unpalatable choice.

In principle, IPv4 scarcity might spur transition to v6. But here too, individual incentives oppose transition. In the short run, a network can use NAT to let a single v4 address serve multiple computers, as discussed in Section 0. At some cost for internal renumbering, the network can reassign and reuse any unused or underused addresses it may have. Finally, the network may be able to transfer addresses from others – either an official transfer as discussed in Section 0, or a "black market" transfer prohibited by formal policy. In the long run, these workarounds carry high costs: As discussed in Section 0, NAT adds complexity, impedes flexibility, and remains untested at the scale some networks might eventually require. Similarly, underused addresses will eventually become hard to find – so reusing addresses cannot continue indefinitely. But in the short run, these v4 challenges are easier than implementing v6. Thus, facing v4 scarcity, it seems networks will naturally choose to use v4 more intensively – not to move to v6.

The core hindrance to v6 seems to be lack of end-user demand for IPv6, for lack of v6-specific features that users value. Suppose users *wanted* v6 – perhaps to obtain higher-quality Skype calls, faster Bittorrent downloads, or more immersive online video games. Seeking such features, users would pressure their ISPs for v6 connectivity. But at present, no such features exist: v6 offers no clear foundations to support such features, and application developers face an overwhelming incentive to make their best features available to v4-only users. Without user demand, the main proponents of v6 are engineers anticipating future design challenges – a less powerful claim on networks' budgets.

Early experience with IPv6 revealed additional disincentives to its use. For one, even when v6 access works, it is often slower than v4: Fewer networks support v6, so v6 data typically flows less directly, often requiring lengthy "tunnel" detours to bypass v4-only networks [15]. Furthermore, v6 malfunctions can make v6-capable services slower and less reliable than those that support only v4. Even if a web site and user are both v6-capable, their connection will fail if an intervening ISP has not set up v6 or has allowed its equipment to malfunction (a more frequent occurrence with v6 than with v4). Furthermore, consider a user who has accidentally enabled v6, whether by hand or through malfunctioning automatic configurations. (For example, some security software enables v6 in order to secure it – incorrectly telling a user's computer that v6 is ready to use.) Initial measurements indicate that misconfigured-v6 users constitute one third of computers currently attempting to use v6 [18]. When any such user attempts to browse a v6-capable site, the user's computer will chose v6 transit – a request which will fail or endure a lengthy timeout before reverting to v4. Meanwhile, affected users can browse all v4-only sites as usual, without delay. As a result, a site *suffers* from enabling v6 – incurring costs such as lost users, slow load times, and user complaints. These incentives have led some early v6-capable sites to *remove* v6 addresses from their servers [19].

Available data confirms the limited deployment of IPv6 to date. For example, Packet Clearing House reports that 78% of Internet exchange points lack v6 support [20] – preventing participating networks from using those exchanges to transfer v6 traffic. Internet routers hold nearly 200 times as many v4 routes as v6 routes [21]. Technical professionals at the APNIC web site still favor v4 by a ratio of 500 to 1 [21].

In short, v6 deployment remains slow and continues to lack the network effects that accelerated deployment of successful Internet standards. It seems unreasonable to expect v6 to succeed on any particular timetable – particularly because self-interest may lead rational networks to prefer v4 (including NAT) over v6 in the short run.

6 A Market Mechanism for Transfer and Reuse of IPv4 Addresses

Even if new IPv4 addresses become unavailable from IANA and RIRs, v4 addresses will continue to be held by existing networks. Some networks may have more than they need due to shrinkage, overoptimistic growth forecasts, or address-saving technologies (*e.g.* v6 or v4 NAT). Other networks may have received abundant "legacy" addresses decades ago. These sources could provide at least temporary relief to v4 scarcity.

6.1 The Historic Prohibition on IPv4 Transfers

Historically, IP addresses have not been transferable between networks. If an operator no longer needs some addresses, the operator may only return the addresses to its

RIR. When one company acquires another, addresses may move with the acquired company [22]. But RIRs have prohibited transactions solely to transfer IP addresses.

6.2 Paid Transfers to Achieve Allocative Efficiency

IPv4 scarcity will create strong incentives for transfers. Some operators will have much less address space than they need. (Consider new operators who receive no addresses prior to exhaustion of available v4 addresses from RIRs.) Conversely, other operators will have more than they need, as discussed above. With transfers, those who most highly value addresses would be likely to obtain addresses from those who can provide addresses at lowest cost – creating surplus from the difference between the parties' valuations.

Consider implications for users who cannot readily switch to v6 – perhaps due to custom software that requires v4, applications incompatible with NAT, unusually costly or busy IT staff, or strong customer or partner preferences. Without transfers, these users would be forced to move to v6 promptly, despite their high costs of transition. In contrast, v4 transfers let these users pay *others* to switch (or otherwise vacate addresses) instead.

Conversely, paid transfers of IPv4 addresses create an incentive for networks to offer addresses for others' use. Under current policies, networks have little incentive to return excess v4 resources: The addresses might be useful or valuable in the future, and a network would forfeit such value if it simply returned addresses to its RIR. In contrast, a paid transfer system pays networks for their unneeded resources – thereby encouraging returns, and rewarding networks which vacate v4 space for use by others.

Experience in other markets indicates that trading resource rights can achieve large efficiency gains. For example, tradable pollution rights reportedly reduced pollution for 55% lower cost than ordering across-the-board cuts by all firms [23] thanks to variation in firms' costs of abatement. Networks feature similar variation in their initial address allocations, in the suitability of transition technologies to serve their requirements, and in their staff and equipment costs of migration. Through transfers, networks with low transition costs can move to v6 first – at lower cost than transitions in random order.

But paid transfers threaten other aspects of addressing policy. Subsequent sections consider possible restrictions on transfers to achieve allocative efficiency while avoiding apparent negative externalities.

6.3 Hybrid Markets to Prevent Speculation

Experience in other markets reveals a risk of speculation, manipulation, and other market anomalies. Cornering the market would be costly [24] and probably ill-advised, but even the risk of such disruption worries those whose businesses would be affected [24]. These concerns invite evaluation of market rules to discourage speculation.

One possibility comes in the form of a *hybrid market*, requiring that each recipient satisfy two separate tests to receive addresses. First, the recipient would have to satisfy a substantive examination in which an RIR verifies the recipient's eligibility. Then, the recipient would need to pay to receive addresses from a provider – with a market mechanism serving both to set price and to find and motivate counterparts.

Substantive review by an RIR would prevent speculators from participating in the market, for speculators would fail an RIR's assessment of need. Moreover, consistent with current practice, an RIR could confirm the *amount* of each network's need – preventing networks from partnering with speculators or from seeking excess addresses in an attempt to corner the market, increase price, or disrupt competitors.

In other contexts, detailed verification might be rejected as overly costly. But RIRs have long performed such review efficiently and at low cost [25].

In other contexts, detailed verification might raise significant concerns of regulatory error. But RIRs have decades of experience evaluating requests, including experience embodied in staff, software, and procedures. Moreover, incorrect authorization grants leave the system little worse off than a process that omits need-based review.

Prohibiting speculation rejects the price discovery benefits of speculation and arbitrage. As a result, prices would be slower to adjust to new information. But network operators seem to consider this loss acceptable in light of the associated benefits [26].

6.4 Preventing Unreasonable Growth of the Routing Table

Paid transfers are in tension with growth of Internet routing tables, and certain transfers create negative externalities that policy might seek to address. I begin by examining key routing characteristics. I then consider and compare policy responses.

6.4.1 The Routing System and Routing Externalities

The Internet's routing system determines how best to transport data between networks. Ordinarily, network addresses are aggregated hierarchically: Data destined for any address in a grouping can be sent to that grouping, without requiring that distant devices know the details of a faraway group. Aggregation offers large efficiency benefits: Although the Internet connects hundreds of millions of devices, routing decisions are several orders of magnitude smaller. Yet routing remains challenging: The Internet's broad reach and complicated structure yield more than 240,000 entries in a full routing table [27]. Moreover, a typical router must forward hundreds of thousands of packets per second, and routes change frequently due to network disruptions and reconfigurations.

Routing table growth imposes substantial costs. Assessing router cost and capacity, network engineer Bill Herrin estimates a cost of $0.04 per route per router per year [28]. Summing over an estimated 150,000 affected routers, each new route costs the Internet community $6,200 per year. Moreover, if routing tables grow rapidly, ISPs might have to replace routers more often than expected – yielding costs above Herrin's projection. Sharply increased growth could even exceed the capabilities of routers reasonably available in the short run [29].

Few incentives currently constrain growth of the routing table. No central authority has meaningful control over route announcements. Nor can ISPs safely reject unwanted route announcements: Route rejections may prevent an ISP's customers from reaching a remote network – prompting customer complaints and unexpected costs. Instead, routing largely follows from address policy: So long as most networks receive large blocks of contiguous addresses, addresses can be aggregated to avoid unnecessary growth in the routing table.

6.4.2 The Effect of Transfers on Route Disaggregation

Paid transfers threaten routing because transfers invite recipients to receive small blocks of addresses from multiple providers – requiring correspondingly many routing table entries. For example, suppose a network needs 2^{16} addresses. If the network obtains 2^{16} contiguous addresses, others' routers can add just a single routing entry. But if the network instead obtains eight noncontiguous blocks of 2^{13}, eight routing entries will be required. If a network considers only its self-interest, it will choose the eight 2^{13} blocks over the single 2^{16} any time the former costs less – imposing a negative externality through extra routing costs.

To address this externality, restrictions could bind either side of the market – regulating address providers, address recipients, or both. In the following sections, I suggest that limiting recipients may be the best choice.

6.4.3 Prohibiting Disaggregation by Address Providers

Policy could stop disaggregation at its source by disallowing or severely limiting disaggregation *by address providers*. In a complete prohibition on disaggregation, a provider might have to transfer all its addresses to a single recipient – not to multiple smaller recipients. By keeping blocks intact, this approach would make it unlikely that transfers would require additional routing entries. Indeed, if a network had to give up its entire space (keeping none for its own ongoing use), a full prohibition on disaggregation might allow transfers with no effect on routing at all.

However, restrictions on provider disaggregation blunt the benefits anticipated from transfers. It appears that future networks will seek smaller blocks of v4 space than typical current allocations. (For one, many transfers are expected to come from legacy holders, whose allocations are often very large. Furthermore, recipients are likely to use v4 space in ways that need only small blocks of v4 space, *e.g.* to host servers or to provide interfaces to NAT gateways.) If policy substantially restricts disaggregation by address providers, there would likely be a glut of large blocks, yet an inadequate supply of small blocks – preventing many networks from sharing the large resources embodied in the large blocks.

In principle, policy can allow limited disaggregation to increase supply of small blocks. For example, providers could be permitted to disaggregate by, *e.g.* a factor of four. But setting cutoffs adds significant complexity, and it would be difficult to determine optimal values. Furthermore, policy changes invite gaming and delay as providers anticipate possible changes and try to optimize their decisions accordingly.

6.4.4 Regulating Recipients through a "Full-Fill Rule"

Alternatively, policy could seek to prevent unreasonable routing table growth by limiting disaggregation requested *by address recipients*. Suppose an RIR's review qualifies a network to receive some specific quantity of v4 addresses. If the network instead requests multiple smaller blocks that sum to the authorized amount, the RIR would reject the request. After all, the network's need could have been satisfied by a single block, reducing the routing burden imposed on others. Recent ARIN discussions call this approach the "full-fill rule" – requiring that a network satisfy its entire need (for some designated period, *e.g.* six months) with a single transfer [30].

Combining the full-fill rule with permissive disaggregation by address providers creates incentives against unreasonable disaggregation. In particular, these rules guarantee that prices will be convex: large blocks will cost at least as much, pro rata, as small blocks. See proof in the Appendix. With convex prices, address providers prefer to transfer their space in as few, large transactions as possible – yielding convexly greater revenue as well as lower transaction costs. Thus, providers attempt to keep blocks intact – a decentralized approach to the negative externality of unreasonable disaggregation.

Moreover, this combination of rules grants v4 space to those who value it most highly. Suppose multiple small recipients are willing to pay more for their joint use of a given block of addresses – exceeding any single large recipient seeking the same total quantity. Then the large recipient is not the highest and best use of those addresses. By allocating the addresses to the various small recipients, the provider can create more value in the sense of Section 0. Arguably, such transfers should be permitted: Disaggregation to connect these new networks is *appropriate* disaggregation which usefully enlarges the Internet; it is not the unreasonable disaggregation policy seeks to prevent. The full-fill rule exactly achieves this result, whereas limiting disaggregation by address providers tends to impede even these desirable instances of disaggregation.

6.5 Avoiding Transferring Addresses from Poor Regions to Rich Regions

Paid transfer of IPv4 addresses could include transfers between regions. For example, a network in a low-income country might find it profitable to transfer its addresses to a network in a high-income country. From one perspective, this is allocatively efficient in the sense of Section 0: The low-income network prefers money over its v4 addresses, while the high-income network needs the addresses more than the money. Furthermore, the low-income network can use the payment to improve other aspects of its service or, via payments to its owners, to otherwise invest in the local economy. So some may conclude that inter-region transfers are laudable and, in any event, ought not be prevented.

v4 transfers may entail important dynamic consequences. If a network implements NAT address sharing rather than globally unique v4 addresses, it may be hindered in the use of new or innovative applications. Some may be troubled by the prospect of such obstacles disproportionately affecting low-income countries. (Perhaps such countries would later suffer second-rate Internet access, further limiting development.) Those who dislike NAT can move to v6, escaping NAT's restrictions. But if v6 expertise and equipment are particularly scarce or costly in low-income countries, v6 may offer little assistance in the short run.

A natural policy response would allow transfers *within* each RIR, but disallow transfers *between* regions. Because substantial wealth variation occurs between RIR service areas, this restriction would sharply reduce likely address transfer from poor countries to rich. That said, a prohibition on inter-region transfers requires careful evaluation. For one, the restriction would prohibit some exchanges that are allocatively efficient in the sense of Section 0 – harming both the would-be recipient and would-be provider. For another, the restriction invites circumvention: Large networks might begin to operate (or claim to operate) within each region so they can receive addresses everywhere. Finally, because the largest share of underutilized v4 resources appear to reside in North America (reflecting early Internet usage and generous address allocations to early users), such a restriction might actually keep prices *lower* in North America than elsewhere. To date, there is no clear consensus on this restriction.

7 The Decision at Hand

Once RIRs can no longer grant more IPv4 addresses, networks will face an unavoidable choice: Share addresses through NAT gateways? Deploy v6 immediately? Pay to receive v4 addresses from others? With long-term tradeoffs and significant uncertainty, the decision is challenging.

A market mechanism for v4 addresses appears to offer important benefits. By putting a positive price on existing addresses, paid transfers would show existing networks how much their addresses are worth to others – giving those networks a direct incentive to make the addresses available to others if they can do so cost-effectively, and offering those networks a financial bonus to spur their migration to v6. Meanwhile, by transferring addresses to networks that cannot easily reduce demand for v4, paid transfers can reduce total system costs – helping the Internet continue to grow. v6 may remain necessary in the long run, but in the short run v4 transfers can help both to mitigate the worst effects of v4 scarcity, and to build the incentives necessary for transition to v6.

References

1. Huston, G.: IPv4 Address Report (May 21, 2008),
   http://www.potaroo.net/tools/ipv4/
2. Hain, T.: A Pragmatic Report on IPv4 Address Space Consumption. The Internet Protocol Journal 8(3) (September 2005)
3. Huston, G., Kosters, M.: Update on Resource Certification. CIDR Report (March 2008)
4. Evans, K.: OMB Memorandum M-05-22 (August 2, 2005)
5. Federal Government Transition IP Version 4 to IP Version 6 – Frequent Asked Questions (February 15, 2006)
6. Nakamura, T.: IPv6 Deployment Status in Japan. Presentation at APNIC 23
7. Saltzer, J.H., Reed, D.P., Clark, D.D.: End-to-End Arguments in System Design. ACM Transactions on Computer Systems 2(4), 277–288 (1984)
8. Solensky, F.: Continued Internet Growth. In: Proceedings of the 18th IETF (August 1990)
9. Internet Protocol, Version 6 (IPv6) Specification. RFC 2460
10. Durand, A.: Issues with NAT-PT DNS ALG in RFC 2766. Internet Draft (January 29, 2003)
11. Network Address Translation - Protocol Translation. RFC 2766
12. Reasons to Move NAT-PT to Historic Status. RFC 4966
13. APRICOT 2008 – Lessons Learned. IPv4 / IPv6 Operational Information Collection
14. Piscitello, D.: IPv6 Support Among Commercial Firewalls. Presentation at ARIN XX
15. Domingues, M., et al.: Is Global IPv6 Deployment on Track? FCCN (October 2007)
16. IIJ Dual-Stack Agreement, http://www.iij.ad.jp/en/development/tech/IPv6/dual/index.html
17. Bytemark Hosting SLA,
    http://www.bytemark.co.uk/page/Live/company/terms

18. Your.org. Working vs. Broken v6 Clients
19. Toyama, K., et al.: Clear and Present Danger of IPv6: IPv6/IPv4 Fallback. NANOG 39
20. Packet Clearing House Report on Distribution of IPv6-Enabled IXPs (January 24, 2009)
21. Geoff, H., Michaelson, G.: IPv6 Deployment: Just Where Are We? ISP Column (April 2008)
22. ARIN Number Resource Policy Manual. Section 8.1
23. Cramton, P.: A Review of Markets for Clean Air. Journal of Economic Literature 38 (September 2000)
24. Jarrow, R.: Market Manipulation, Bubbles, Corners, and Short Squeezes. The Journal of Financial and Quantitative Analysis 27(3), 311–336 (1992)
25. IPv4 Transfer Policy Proposal – ARIN XXI Public Policy Meeting Transcript. ARIN XXI (April 7, 2008)
26. Transfer Policy Proposal Survey – Results. ARIN Advisory Council (August 26, 2008)
27. Huston, G.: Nanogging. ISP Column (November 2007)
28. Herrin, B.: BGP Cost (February 2008)
29. Panel: IP Markets – ARIN XX Public Policy Meeting Draft Transcript,
    http://www.arin.net/meetings/minutes/ARIN_XX/
    ppm1_transcript.html#anchor_7
30. Darte, B.: Emergency Transfer Policy for IPv4 Addresses. ARIN Policy Proposal 2008-6

Appendix: Full-Fill Plus Permissive Disaggregation Guarantees Convex Prices

Claim: Suppose address recipients are bound by the full-fill rule, and suppose providers may disaggregate as they see fit. Then prices are weakly convex. That is, if $P(q)$ is the prevailing market price for a block of size q, then for any Q and for any $a>1$, it must be the case that $P(aQ)>aP(Q)$.

Proof: Suppose not. Then there exists a provider with some quantity Q that could be divided into Q_1 and Q_2 where $Q=Q_1+Q_2$ but $P(Q)<P(Q_1)+P(Q_2)$. If so, the provider would never transfer a Q block intact, but rather would subdivide that Q into smaller blocks Q_1 and Q_2, increasing revenue. So $P(Q)$ cannot be the price of a block of size Q.

The following graph shows the impossibility of concave prices. The transferor would increase revenue by subdividing its Q-sized block into separate blocks of size Q_1 and Q_2. In particular, notice that $P(Q_1)+P(Q_2)>P(Q)=P(Q_1+Q_2)$.

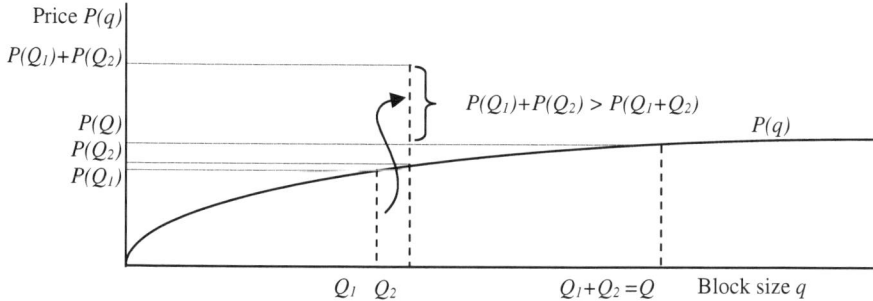

Author Index

Bichler, Martin   1, 93

Capponi, Agostino   4
Charles, Denis   55
Cherubini, Umberto   4
Chickering, Max   55
Conitzer, Vincent   61
Cowgill, Bo   3

Dughmi, Shaddin   89
Dutta, Partha   74

Edelman, Benjamin   95

Farfel, Joseph   61
Friedman, Eric J.   13

Guler, Kemal   93

Halpern, Joseph Y.   13

Jennings, Nicholas R.   74

Kash, Ian A.   13

Lahaie, Sébastien   92
Lee, Sooyeon   40
Lochner, Kevin M.   26

Macbeth, Douglas K.   74
Miralles, Antonio   58
Mullen, Tracy   40

Parkes, David C.   92
Puri, Sidd   55

Reyes H., Gonzalo   25
Rogers, Alex   74
Roughgarden, Tim   89

Said, Maher   87
Sayal, Mehmet   93
Schneider, Stefan   1, 93
Seuken, Sven   55
Shabalin, Pasha   1
Sohn, Jung-woo   40
Stranjak, Armin   74
Sundararajan, Mukund   89

Vytelingum, Perukrishnen   74

Wellman, Michael P.   26
Wolfers, Justin   3

Zitzewitz, Eric   3